PLUMBING
ESSENTIALS

COWLES
Creative Publishing
A Division of Cowles Enthusiast Media, Inc.

BLACK&
DECKER®
HOME TIPS™

Credits

Copyright © 1996
Cowles Creative Publishing, Inc.
Formerly Cy DeCosse Incorporated
5900 Green Oak Drive
Minnetonka, Minnesota 55343
1-800-328-3895
All rights reserved
Printed in U.S.A.

COWLES
Creative Publishing
A Division of Cowles Enthusiast Media, Inc.

President/COO: Nino Tarantino
Executive V.P./Editor-in-Chief: William B. Jones

Created by: The Editors of Cowles Creative Publishing, Inc.,
in cooperation with Black & Decker. is
a trademark of the Black & Decker Corporation and is
used under license.

Printed on American paper by:
 Quebecor Printing
 99 98 97 96 / 5 4 3 2 1

COWLES
Enthusiast Media

President/COO: Philip L. Penny

Books available in this series:

Wiring Essentials
Plumbing Essentials
Carpentry Essentials
Painting Essentials
Flooring Essentials
Landscape Essentials
Masonry Essentials
Door & Window Essentials
Roof System Essentials
Deck Essentials
Porch & Patio Essentials
Built-In Essentials

Contents

The Home Plumbing System

Because most of a plumbing system is hidden inside walls and floors, it may seem to be a complex maze of pipes and fittings. In fact, home plumbing is simple and straightforward. Understanding how home plumbing works is an important first step toward doing routine maintenance and money-saving repairs.

A typical home plumbing system includes three basic parts: a water supply system, fixtures and appliances, and a drain system. These three parts can be seen clearly in the photograph of the cutaway house on the opposite page.

Fresh water enters a home through a main supply line (1). This fresh water source is provided by either a municipal water company or a private underground well. If the source is a municipal supplier, the water passes through a meter (2) that registers the amount of water used. A family of four uses about four hundred gallons of water each day.

Immediately after the main supply enters the house, a branch line splits off (3) and is joined to a hot water heater (4). From the water heater, a hot water line runs parallel to the cold water line to bring the water supply to fixtures and appliances throughout the house. Fixtures include sinks, bathtubs, showers, and laundry tubs. Appliances include water heaters, dishwashers, clothes washers, and water softeners. Toilets and exterior sillcocks are examples of fixtures that require only a cold water line.

The water supply to fixtures and appliances is controlled with faucets and valves. Faucets and valves have moving parts and seals that eventually may wear out or break, but they are easily repaired or replaced.

Waste water then enters the drain system. It first must flow past a trap (5), a U-shaped piece of pipe that holds standing water and prevents sewer gases from entering the home. Every fixture must have a drain trap.

The drain system works entirely by gravity, allowing waste water to flow downhill through a series of large-diameter pipes. These drain pipes are attached to a system of vent pipes. Vent pipes (6) bring fresh air to the drain system, preventing suction that would slow or stop drain water from flowing freely. Vent pipes usually exit the house at a roof vent (7).

All waste water eventually reaches a main waste and vent stack (8). The main stack curves to become a sewer line (9) that exits the house near the foundation. In a municipal system, this sewer line joins a main sewer line located near the street. Where sewer service is not available, waste water empties into a septic system.

Water meter and main shutoff valve are located where the main water supply pipe enters the house. The water meter is the property of your local municipal water company. If the water meter leaks, or if you suspect it is not functioning properly, call your water company for repairs.

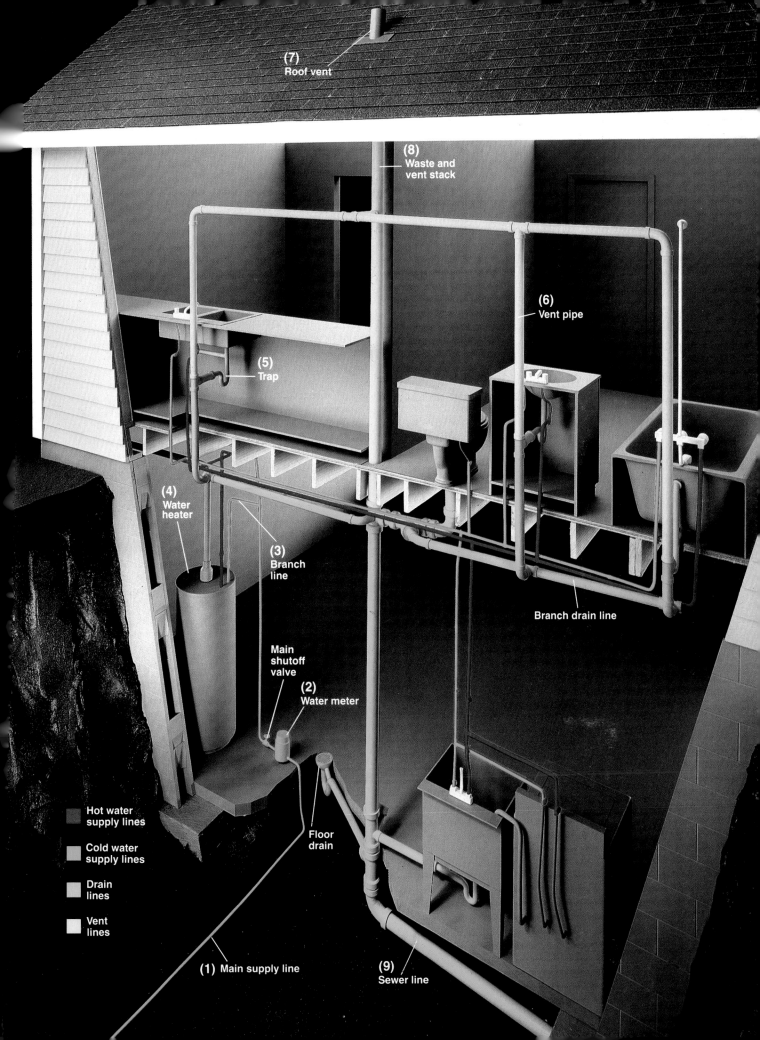

(7) Roof vent

(8) Waste and vent stack

(6) Vent pipe

(5) Trap

Branch drain line

(4) Water heater

(3) Branch line

Main shutoff valve

(2) Water meter

Floor drain

Hot water supply lines

Cold water supply lines

Drain lines

Vent lines

(1) Main supply line

(9) Sewer line

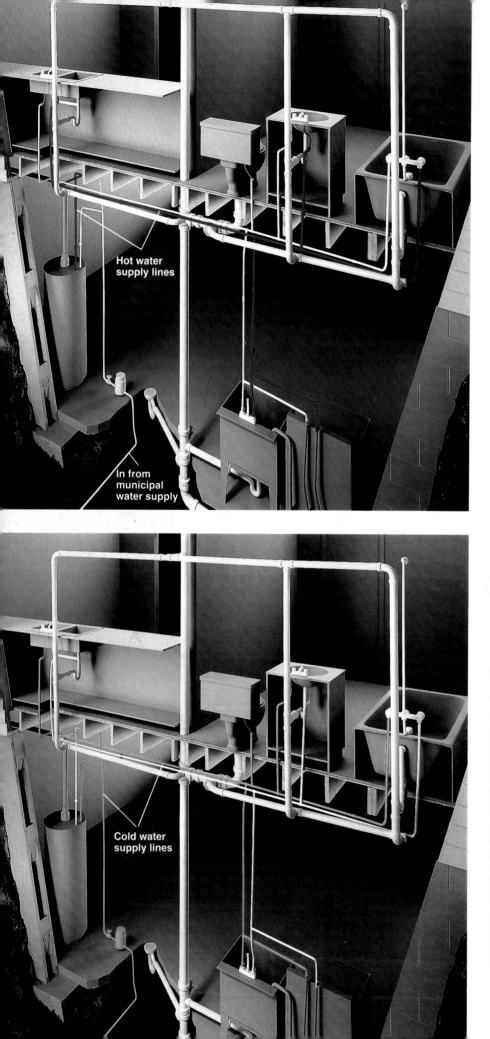

Hot water
supply lines

In from
municipal
water supply

Cold water
supply lines

Water Supply System

Water supply pipes carry hot and cold water throughout a house. In homes built before 1950, the original supply pipes are usually made of galvanized iron. Newer homes have supply pipes made of copper. In some areas of the country, supply pipes made of plastic are gaining acceptance by local plumbing codes.

Water supply pipes are made to withstand the high pressures of the water supply system. They have small diameters, usually ½" to 1", and are joined with strong, watertight fittings. The hot and cold lines run in tandem to all parts of the house. Usually, the supply pipes run inside wall cavities or are strapped to the undersides of floor joists.

Hot and cold water supply pipes are connected to fixtures or appliances. Fixtures include sinks, tubs, and showers. Some fixtures, such as toilets or hose bibs, are supplied only by cold water. Appliances include dishwashers and clothes washers. A refrigerator is an example of an appliance that uses only cold water. Tradition says that hot water supply pipes and faucet handles are found on the left-hand side of a fixture. Cold water is on the right.

Because of high water pressure, leaks are the most common problems for the water supply system. This is especially true of galvanized iron pipe, which has limited resistance to corrosion.

Drain, Waste, Vent System

Drain pipes use gravity to carry waste water away from fixtures, appliances, and other drains. This waste water is carried out of the house to a municipal sewer system or septic tank.

Drain pipes are usually plastic or cast iron. In some older homes, drain pipes may be made of copper or lead. Because they are not part of the supply system, lead drain pipes pose no health hazard. However, lead pipes are no longer manufactured for home plumbing systems.

Drain pipes have diameters ranging from 1¼" to 4". These large diameters allow waste water to pass easily.

Traps are an important part of the drain system. These curved sections of drain pipe hold standing water, and they are usually found near any drain opening. The standing water of a trap prevents sewer gases from backing up into the home. Each time a drain is used, the standing trap water is flushed away and is replaced by new water.

In order to work properly, the drain system requires air. Air allows waste water to flow freely down drain pipes.

To allow air into the drain system, drain pipes are connected to vent pipes. All drain systems must include vents, and the entire system is called the drain, waste, vent (DWV) system. One or more vent stacks, located on the roof, provide the air needed for the DWV system to work.

Vent

Vent lines

Trap

Drain lines

Out to municipal sewer

Tools for Plumbing

Many plumbing projects and repairs can be completed with basic hand tools you probably already own. Adding a few simple plumbing tools will prepare you for all the projects in this book. Specialty tools, such as a cast iron cutter or appliance dolly, are available at rental centers. When buying tools, invest in quality products.

Always care for tools properly. Clean tools after using them, wiping them free of dirt and dust with a soft rag. Prevent rust on metal tools by wiping them with a rag dipped in household oil. If a metal tool gets wet, dry it immediately, and then wipe it with an oiled rag. Keep tool boxes and cabinets organized. Make sure all tools are stored securely.

Caulk gun is designed to hold tubes of caulk or glue. A squeeze handle pushes a steady bead of caulk or glue out of the nozzle.

Flashlight is an indispensable plumber's helper for inspecting pipes and drain openings.

Circuit tester is an important safety device that allows the user to test for live current in an electrical outlet or appliance. Also referred to as *testing for hot wires.*

Ratchet wrench is used for tightening or loosening bolts and nuts. It has interchangeable sockets for adapting to different sized bolts or nuts.

Hacksaw is used for cutting metals. Also can be used for cutting plastic pipes. Has replaceable blades.

Small wire brush has soft brass bristles for cleaning metals without damaging surfaces.

Utility knife has a razor-sharp blade for cutting a wide variety of materials. Useful for trimming ends of plastic pipes. For safety, the utility knife should have a retractable blade.

Cold chisel is used with a *ball peen hammer* to cut or chip ceramic tile, mortar, or hardened metals.

Files are used to smooth the edges of metal, wood, or plastic. The *round file* (top) can be used to remove burrs from the insides of pipes. The *flat file* is used for all general smoothing tasks.

Screwdrivers include the two most common types: the *slotted* (top), and the *phillips*.

Adjustable wrench has a movable jaw that permits the wrench to fit a wide variety of bolt heads or nuts.

Channel-type pliers has a movable handle that allows the jaws to be adjusted for maximum gripping strength. The insides of the jaws are serrated to prevent slipping.

Needlenose pliers has thin jaws for gripping small objects, or for reaching into confined areas.

Putty knife is especially helpful for scraping away old putty or caulk from appliances and fixtures.

Ball peen hammer is made for striking metallic objects, like a *cold chisel*. The head of a ball peen hammer is made to resist chipping.

Wooden mallet is used for striking nonmetallic objects, such as plastic drywall anchors.

Tape measure should have a retractable steel blade at least 16 feet long.

Level is used for setting new appliances and checking the slope of exhaust ducts.

Tubing cutter makes straight, smooth cuts in plastic and copper pipe. A tubing cutter usually has a dull, triangular blade, called a *reaming tip*, for removing burrs from the insides of pipes.

Closet auger is used to clear toilet clogs. It is a slender tube with a crank handle on one end of a flexible auger cable. A special bend in the tube allows the auger to be positioned in the bottom of the toilet bowl. The bend is usually protected with a rubber sleeve to prevent scratching the toilet.

Plastic tubing cutter works like a gardener's pruners to cut flexible plastic (PB) pipes quickly.

Spud wrench is specially designed for removing or tightening large nuts that are 2'' to 4'' in diameter. Hooks on the ends of the wrench grab onto the *lugs* of large nuts for increased leverage.

Plunger clears drain clogs with water and air pressure. The *flanged plunger* (shown) is used for toilet bowls. The flange usually can be folded up into the cup for use as a *standard plunger*. Use a standard plunger to clear clogs in sink, tub, shower, and floor drains.

Hand auger, sometimes called a *snake*, is used to clear clogs in drain lines. A long, flexible steel cable is stored in the disk-shaped crank. A pistol-grip handle allows the user to apply steady pressure on the cable.

Blow bag, sometimes called an *expansion nozzle,* is used to clear drains. It attaches to a garden hose and removes clogs with powerful spurts of water. The blow bag is best used on floor drains.

Propane torch (left) is used for soldering fittings to copper pipes. Light the torch quickly and safely using a **spark lighter** (above).

Power hand tools can make any job faster, easier, and safer. Cordless power tools offer added convenience. Use a cordless ⅜" **power drill** for virtually any drilling task. A cordless **power ratchet** makes it easy to turn small nuts or hex-head bolts. The cordless reversible **power screwdriver** drives a wide variety of screws and fasteners. A **reciprocating saw** uses interchangeable blades to cut wood, metal, or plastic. Thaw frozen pipes fast with a **heat gun.**

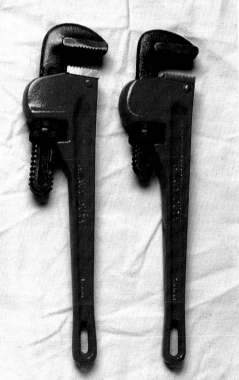

Pipe wrench has a movable jaw that adjusts to fit a variety of pipe diameters. Pipe wrench is used for tightening and loosening pipes, pipe fittings, and large nuts. Two pipe wrenches often are used together to prevent damage to pipes and fittings.

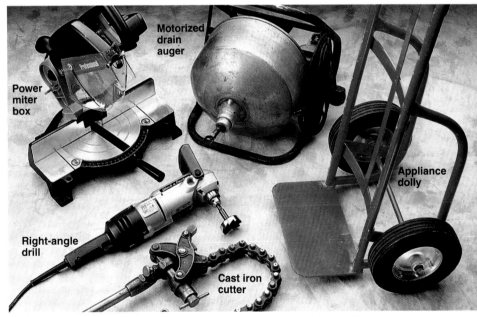

Rental tools may be needed for large jobs and special situations. A **power miter box** makes fast, accurate cuts in a wide variety of materials, including plastic pipes. A **motorized drain auger** clears tree roots from sewer service lines. Use an **appliance dolly** to move heavy objects like water heaters. A **cast iron cutter** is designed to cut tough cast-iron pipes. The **right-angle drill** is useful for drilling holes in hard-to-reach areas.

Plumbing Materials

Check local plumbing code for materials allowed in your area. All diameters specified are the interior diameters (I.D.) of pipes.

Benefits & Characteristics

Cast iron is very strong, but is difficult to cut and fit. Repairs and replacements should be made with plastic pipe, if allowed by local code.

ABS (Acrylonitrile-Butadiene-Styrene) was the first rigid plastic approved for use in home drain systems. Some local plumbing codes now restrict the use of ABS in new installations.

PVC (Poly-Vinyl-Chloride) is a modern rigid plastic that is highly resistant to damage by heat or chemicals. It is the best material for drain-waste-vent pipes.

Galvanized iron is very strong, but gradually will corrode. Not advised for new installation. Because galvanized iron is difficult to cut and fit, large jobs are best left to a professional.

CPVC (Chlorinated-Poly-Vinyl-Chloride) rigid plastic is chemically formulated to withstand the high temperatures and pressures of water supply systems. Pipes and fittings are inexpensive.

PB (Poly-Butylene) flexible plastic is easy to fit. It bends easily around corners and requires fewer fittings than CPVC. Not all local codes have been updated to permit use of PB pipe.

Rigid copper is the best material for water supply pipes. It resists corrosion, and has smooth surfaces that provide good water flow. Soldered copper joints are very durable.

Chromed copper has an attractive shiny surface, and is used in areas where appearance is important. Chromed copper is durable and easy to bend and fit.

Flexible copper tubing is easy to shape, and will withstand a slight frost without rupturing. Flexible copper bends easily around corners, so it requires fewer fittings than rigid copper.

Brass is heavy and durable. **Chromed brass** has an attractive shiny surface, and is used for drain traps where appearance is important.

Common Uses	Lengths	Diameters	Fitting Methods	Tools Used for Cutting
Main drain-waste-vent pipes	5 ft., 10 ft.	3″, 4″	Joined with banded neoprene couplings	Cast iron cutter or hacksaw
Drain & vent pipes; drain traps	10 ft., 20 ft.; or sold by linear ft.	1½″, 2″, 3″, 4″	Joined with solvent glue and plastic fittings	Tubing cutter, miter box, or hacksaw
Drain & vent pipes; drain traps	10 ft., 20 ft.; or sold by linear ft.	1½″, 2″, 3″, 4″	Joined with solvent glue and plastic fittings	Tubing cutter, miter box, or hacksaw
Drains; hot & cold water supply pipes	1″ to 1-ft. nipples; custom lengths up to 20 ft.	½″, ¾″, 1″, 1½″, 2″	Joined with galvanized threaded fittings	Hacksaw or reciprocating saw
Hot & cold water supply pipes	10 ft.	⅜″, ½″, ¾″, 1″	Joined with solvent glue and plastic fittings, or with grip fittings	Tubing cutter, miter box, or hacksaw
Hot & cold water supply, where allowed by code	25-ft., 100-ft. coils; or sold by linear ft.	⅜″, ½″, ¾″	Joined with plastic grip fittings	Flexible plastic tubing cutter, sharp knife, or miter box
Hot & cold water supply pipes	10 ft., 20 ft.; or sold by linear ft.	⅜″, ½″, ¾″, 1″	Joined with metal solder or compression fittings	Tubing cutter, hacksaw, or jig saw
Supply tubing for plumbing fixtures	12″, 20″, 30″	⅜″	Joined with brass compression fittings	Tubing cutter or hacksaw
Gas tubing; hot & cold water supply tubing	30-ft., 60-ft. coils; or sold by linear ft.	¼″, ⅜″, ½″, ¾″, 1″	Joined with brass flare fittings, compression fittings, or metal solder	Tubing cutter or hacksaw
Valves & shutoffs; chromed drain traps	Lengths vary	¼″, ½″, ¾″; for drain traps: 1¼″, 1½″	Joined with compression fittings, or with metal solder	Tubing cutter, hacksaw, or reciprocating saw

Water Supply Fittings

Copper	Galvanized iron	CPVC

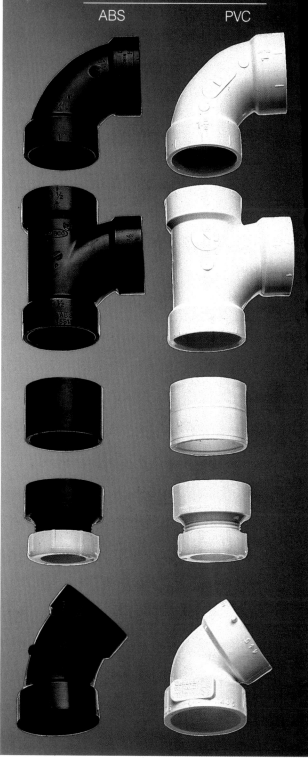

Drain-Waste-Vent Fittings

ABS	PVC

90° elbows are used to make right-angle bends in a pipe run. Drain-waste-vent (DWV) elbows are curved to prevent debris from being trapped in the bend.

T-fittings are used to connect branch lines in water supply and drain-waste-vent systems. A T-fitting used in a DWV system is called a "waste-T" or "sanitary T."

Couplings are used to join two straight pipes. Special transition fittings (page opposite) are used to join two pipes that are made from different materials.

Reducers connect pipes of different diameters. Reducing T-fittings and elbows are also available.

45° elbows are used to make gradual bends in a pipe run. Elbows are also available with 60° and 72° bends.

Plumbing Fittings

Plumbing fittings come in different shapes to let you form branch lines, change the direction of a pipe run, or connect pipes of different sizes. Transition fittings are used to connect pipes and fixtures that are made from different materials (page opposite). Fittings come in many sizes, but the basic shapes are standard to all metal and plastic pipes. In general, fittings used to connect drain pipes have gradual bends for a smooth flow of drain water.

How to Use Transition Fittings

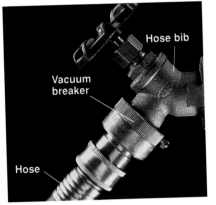

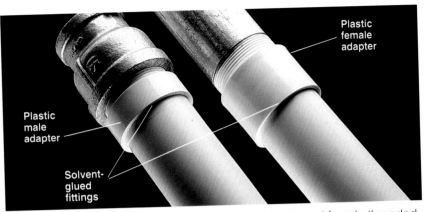

A vacuum breaker is required to connect a hose to a threaded hose bib. It prevents water backing up into the supply pipes.

Connect plastic to threaded metal pipes with male and female threaded adapters. Plastic adapter is solvent-glued to plastic pipe. Threads of pipe should be wrapped with Teflon™ tape. Metal pipe is then screwed directly to the adapter.

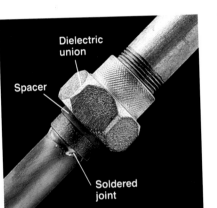

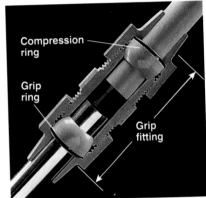

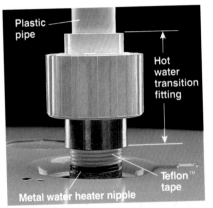

Connect copper to galvanized iron with a dielectric union. Union is threaded onto iron pipe, and is soldered to copper pipe. A dielectric union has plastic spacer that prevents corrosion caused by electro-chemical reaction between metals.

Connect plastic to copper with a grip fitting. Each side of the fitting (shown in cutaway) contains a narrow grip ring and a plastic compression ring (or rubber O-ring) that forms the seal.

Connect metal hot water pipe to plastic with a hot water transition fitting that prevents leaks caused by different expansion rates of materials. Metal pipe threads are wrapped with Teflon™ tape. Plastic pipe is solvent-glued to fitting.

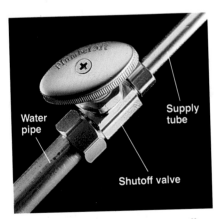

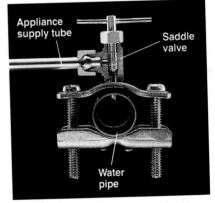

Connect a water pipe to any fixture supply tube, using a shutoff valve (pages 54 to 55).

Connect any supply tube to a fixture tailpiece with a coupling nut. Coupling nut seals the bell-shaped end of supply tube against the fixture tailpiece.

Connect appliance supply tube to copper pipe with a saddle valve (page 57). Saddle valve (shown in cutaway) often is used to connect a refrigerator icemaker.

Working with Copper

Copper is the ideal material for water supply pipes. It resists corrosion and has smooth surfaces that provide good water flow. Copper pipes are available in several diameters (page 13) but most home water supply systems use ½" or ¾" pipe. Copper pipe is manufactured in rigid and flexible forms.

Rigid copper, sometimes called hard copper, is approved for home water supply systems by all local codes. It comes in three wall-thickness grades: Types M, L, and K. Type M is the thinnest, the least expensive, and a good choice for do-it-yourself home plumbing.

Rigid Type L usually is required by codes for commercial plumbing systems. Because it is strong and solders easily, Type L may be preferred by some professional plumbers and do-it-yourselfers for home use. Type K has the heaviest wall thickness, and is used most often for underground water service lines.

Flexible copper, also called soft copper, comes in two wall-thickness grades: Types L and K. Both are approved for most home water supply systems, although flexible Type L copper is used primarily for gas service lines. Because it is bendable and will resist a mild frost, Type L may be installed as part of a water supply system in unheated indoor areas, like crawl spaces. Type K is used for underground water service lines.

A third form of copper, called DWV, is used for drain systems. Because most codes now allow low-cost plastic pipes for drain systems, DWV copper is seldom used.

Copper pipes are connected with soldered, compression, or flare fittings (see chart below). Always follow your local code for the correct types of pipes and fittings allowed in your area.

Soldered fittings, also called sweat fittings, often are used to join copper pipes. Correctly soldered fittings (pages 18 to 22), are strong and trouble-free. Copper pipe can also be joined with compression fittings (pages 24 to 25) or flare fittings. See chart below.

Copper Pipe & Fitting Chart

	Rigid Copper			Flexible Copper		
Fitting Method	**Type M**	**Type L**	**Type K**	**Type L**	**Type K**	**General Comments**
Soldered	yes	yes	yes	yes	yes	Inexpensive, strong, and trouble-free fitting method. Requires some skill.
Compression	yes	not recommended		yes	yes	Easy to use. Allows pipes or fixtures to be repaired or replaced readily. More expensive than solder. Best used on flexible copper.
Flare	no	no	no	yes	yes	Use only with flexible copper pipes. Usually used as a gas-line fitting. Requires some skill.

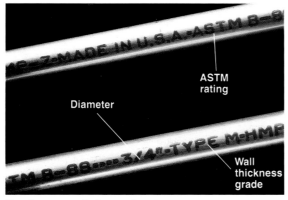

Grade stamp information includes pipe diameter, the wall-thickness grade, and a stamp of approval from the ASTM (American Society for Testing and Materials). Type M pipe is identified by red lettering, Type L by blue lettering.

Bend flexible copper pipe with a coil-spring tubing bender to avoid kinks. Select a bender that matches the outside diameter of the pipe. Slip bender over pipe using a twisting motion. Bend pipe slowly until it reaches the correct angle, but not more than 90°.

Specialty tools & materials for working with copper include: flaring tool (A), emery cloth (B), coil-spring tubing bender (C), pipe joint compound (D), self-cleaning soldering paste (flux) (E), lead-free solder (F), wire brush (G), flux brush (H), compression fitting (I), flare fitting (J).

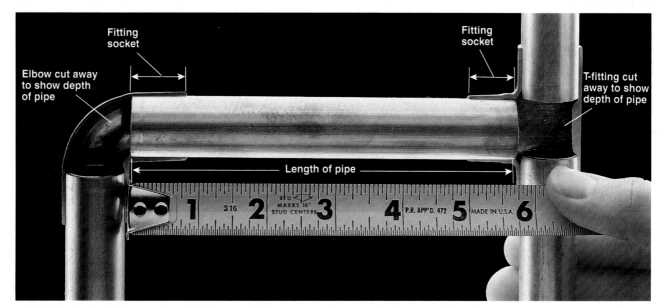

Find length of copper pipe needed by measuring between the bottom of the copper fitting sockets (fittings shown in cutaway). Mark length on the pipe with a felt-tipped pen.

Cutting & Soldering Copper

The best way to cut rigid and flexible copper pipe is with a tubing cutter. A tubing cutter makes a smooth, straight cut, an important first step toward making a watertight joint. Remove any metal burrs on the cut edges with a reaming tool or round file.

Copper can be cut with a hacksaw. A hacksaw is useful in tight areas where a tubing cutter will not fit. Take care to make a smooth, straight cut when cutting with a hacksaw.

A soldered pipe joint, also called a sweated joint, is made by heating a copper or brass fitting with a propane torch until the fitting is just hot enough to melt metal solder. The heat draws the solder into the gap between the fitting and pipe to form a watertight seal. A fitting that is overheated or unevenly heated will not draw in solder. Copper pipes and fittings must be clean and dry to form a watertight seal.

Protect wood from heat of the torch flame while soldering, using a double layer (two 18" × 18" pieces) of 26-gauge sheet metal. Buy sheet metal at hardware stores or building supply centers, and keep it to use with all soldering projects.

Everything You Need:

Tools: tubing cutter with reaming tip (or hacksaw and round file), wire brush, flux brush, propane torch, spark lighter (or matches), adjustable wrench, channel-type pliers.

Materials: copper pipe, copper fittings, emery cloth, soldering paste (flux), sheet metal, lead-free solder, rag.

Soldering Tips

Use caution when soldering copper. Pipes and fittings become very hot and must be allowed to cool before handling.

Keep joint dry when soldering existing water pipes by plugging the pipe with bread. Bread absorbs moisture that may ruin the soldering process and cause pinhole leaks. The bread dissolves when water is turned back on.

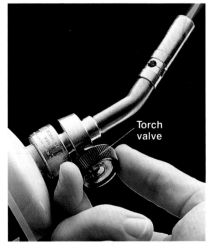

Torch valve

Prevent accidents by shutting off propane torch immediately after use. Make sure valve is closed completely.

How to Cut Rigid & Flexible Copper Pipe

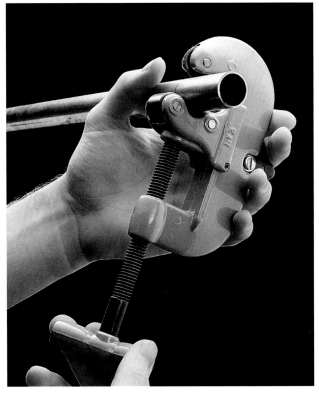

1 Place tubing cutter over the pipe and tighten the handle so that pipe rests on both rollers, and cutting wheel is on marked line.

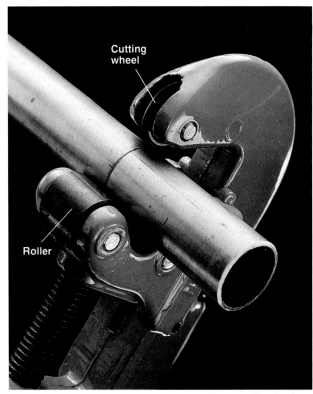

2 Turn tubing cutter one rotation so that cutting wheel scores a continuous straight line around the pipe.

3 Rotate the cutter in the opposite direction, tightening the handle slightly after every two rotations, until cut is complete.

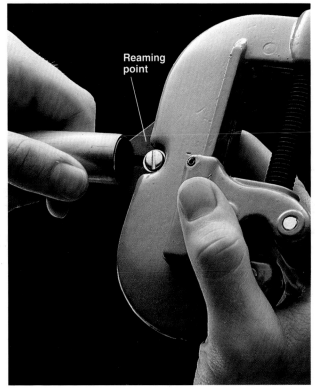

4 Remove sharp metal burrs from inside edge of the cut pipe, using the reaming point on the tubing cutter, or a round file.

How to Solder Copper Pipes & Fittings

1 Clean end of each pipe by sanding with emery cloth. Ends must be free of dirt and grease to ensure that the solder forms a good seal.

2 Clean inside of each fitting by scouring with a wire brush or emery cloth.

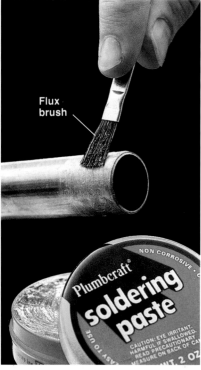

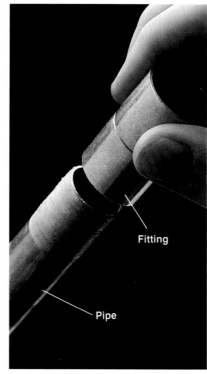

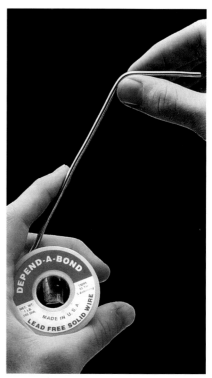

3 Apply a thin layer of soldering paste (flux) to the end of each pipe and to the inside of each fitting, using a flux brush. Soldering paste should cover about 1" of pipe end.

4 Assemble each joint by inserting the pipe into fitting so it is tight against the bottom of the fitting sockets. Twist each fitting slightly to spread soldering paste.

5 Prepare the wire solder by unwinding 8" to 10" of wire from spool. Bend the first 2" of the wire to a 90° angle.

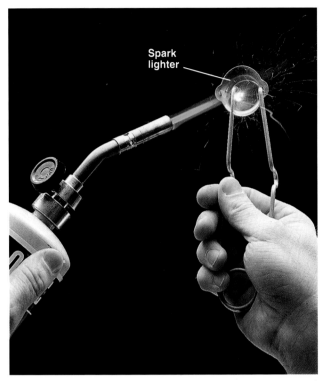

6 Light propane torch by opening valve and striking a spark lighter or a match next to the torch nozzle until the gas ignites.

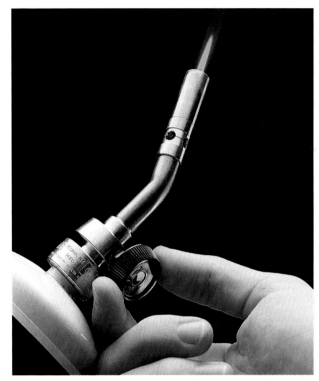

7 Adjust the torch valve until the inner portion of the flame is 1" to 2" long.

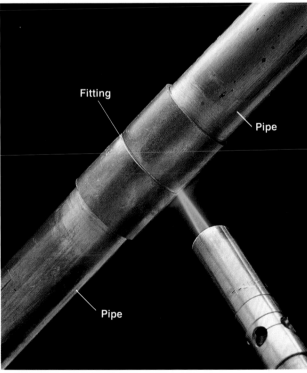

8 Hold flame tip against middle of fitting for 4 to 5 seconds, until soldering paste begins to sizzle.

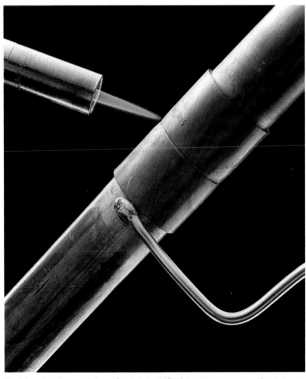

9 Heat other side of copper fitting to ensure that heat is distributed evenly. Touch solder to pipe. If solder melts, pipe is ready to be soldered.

(continued next page)

How to Solder Copper Pipes & Fittings (continued)

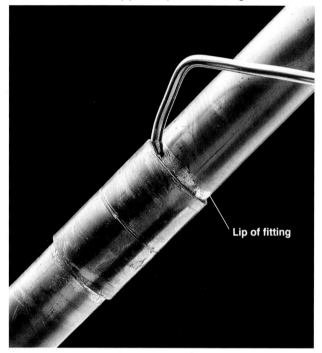

Lip of fitting

10 When pipe is hot enough to melt solder, remove torch and quickly push ½" to ¾" of solder into each joint. Capillary action fills joint with liquid solder. A correctly soldered joint should show a thin bead of solder around the lip of the fitting.

11 Wipe away excess solder with a dry rag. **Caution: pipes will be hot.** When all joints have cooled, turn on water and check for leaks. If joint leaks, drain pipes, apply additional soldering paste to rim of joint, and resolder.

How to Solder Brass Valves

1 Remove the valve stem with an adjustable wrench. Removing the stem prevents heat damage to rubber or plastic stem parts while soldering. Prepare the copper pipes (page 20) and assemble joints.

2 Light propane torch (page 21). Heat body of valve, moving flame to distribute heat evenly. Brass is denser than copper, so it requires more heating time before joints will draw solder. Apply solder (pages 20 to 22). Let metal cool, then reassemble valve.

How to Take Apart Soldered Joints

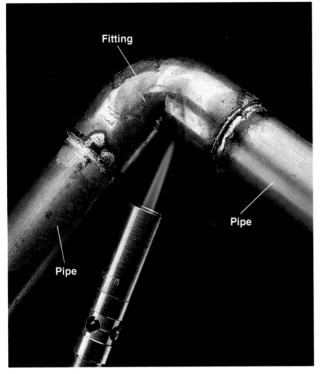

Fitting

Pipe

Pipe

1 Turn off the water (page 4) and drain the pipes by opening the highest and lowest faucets in the house. Light propane torch (page 21). Hold flame tip to the fitting until the solder becomes shiny and begins to melt.

2 Use channel-type pliers to separate the pipes from the fitting.

3 Remove old solder by heating ends of pipe with propane torch. Use dry rag to wipe away melted solder quickly. **Caution: pipes will be hot.**

4 Use emery cloth to polish ends of pipe down to bare metal. Never reuse old fittings.

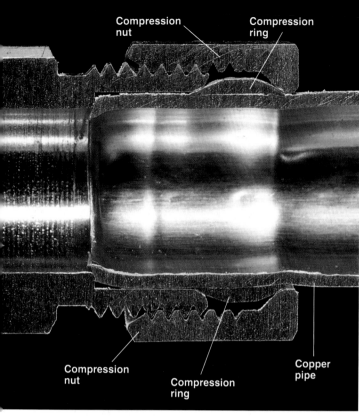

Compression nut   Compression ring

Compression nut   Compression ring   Copper pipe

Compression fitting (shown in cutaway) shows how threaded compression nut forms seal by forcing the compression ring against the copper pipe. Compression ring is covered with pipe joint compound before assembling to ensure a perfect seal.

Using Compression Fittings

Compression fittings are used to make connections that may need to be taken apart. Compression fittings are easy to disconnect, and often are used to install supply tubes and fixture shutoff valves (pages 54 to 55, and sequence below). Use compression fittings in places where it is unsafe or difficult to solder, such as in a crawl space.

Compression fittings are used most often with flexible copper pipe. Flexible copper is soft enough to allow the compression ring to seat snugly, creating a watertight seal. Compression fittings also may be used to make connections with Type M rigid copper pipe. See the chart on page 16.

Everything You Need:

Tools: felt-tipped pen, tubing cutter or hacksaw, adjustable wrenches.

Materials: brass compression fittings, pipe joint compound.

How to Attach Supply Tubes to Fixture Shutoff Valves with Compression Fittings

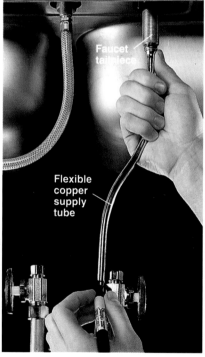

Faucet tailpiece

Flexible copper supply tube

1 Bend flexible copper supply tube, and mark to length. Include ½" for portion that will fit inside valve. Cut tube (page 19).

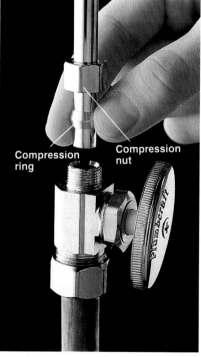

Compression ring   Compression nut

2 Slide the compression nut and then the compression ring over end of pipe. Threads of nut should face the valve.

3 Apply a layer of pipe joint compound over the compression ring. Joint compound helps ensure a watertight seal.

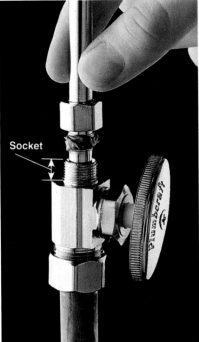

4 Insert end of pipe into fitting so it fits flush against bottom of fitting socket.

Socket

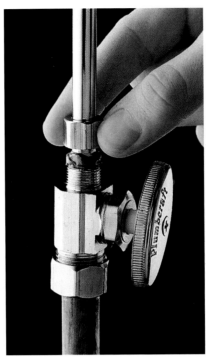

5 Slide compression ring and nut against threads of valve. Hand-tighten nut onto valve.

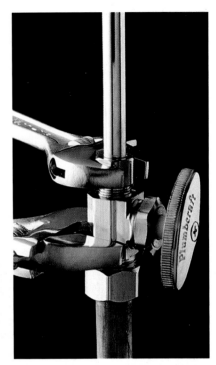

6 Tighten compression nut with adjustable wrenches. Do not overtighten. Turn on water and watch for leaks. If fitting leaks, tighten nut gently.

How to Join Two Copper Pipes with a Compression Union Fitting

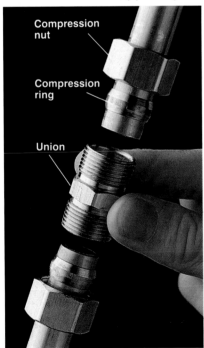

Compression nut

Compression ring

Union

1 Slide compression nuts and rings over ends of pipes. Place threaded union between pipes.

2 Apply a layer of pipe joint compound to compression rings, then screw compression nuts onto threaded union.

3 Hold center of union fitting with an adjustable wrench, and use another wrench to tighten each compression nut one complete turn. Turn on water. If fitting leaks, tighten nuts gently.

Working with Plastics

Plastic pipes can be joined to existing iron or copper pipes using transition fittings (page 15), but different types of plastic should not be joined. For example, if your drain pipes are ABS plastic, use only ABS pipes and fittings when making repairs and replacements.

Plastics pipes are available in rigid and flexible forms. Rigid plastics include ABS (Acrylonitrile-Butadiene-Styrene), PVC (Poly-Vinyl-Chloride), and CPVC (Chlorinated-Poly-Vinyl-Chloride). The most commonly used flexible plastic is PB (Poly-Butylene).

ABS and PVC are used in drain systems. PVC is a newer form of plastic that resists chemical damage and heat better than ABS. It is approved for above-ground use by all plumbing codes. However, some codes still require cast-iron pipe for main drains that run under concrete slabs.

CPVC and PB are used in water supply systems. Rigid CPVC pipe and fittings are less expensive than PB, but flexible PB pipe is a good choice in cramped locations, because it bends easily and requires fewer fittings.

Plastic pipes can be joined to existing iron or copper pipes using transition fittings (page 15), but different types of plastic should not be joined. For example, if your drain pipes are ABS plastic, use only ABS pipes and fittings when making repairs and replacements.

Prolonged exposure to sunlight eventually can weaken plastic plumbing pipe, so plastics should not be installed or stored in areas that receive constant direct sunlight.

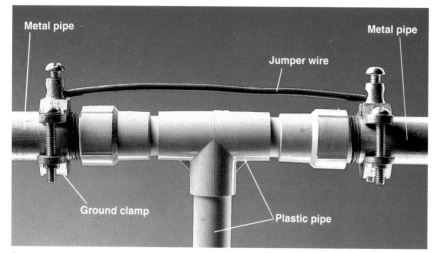

Caution: Your home electrical system could be grounded through metal water pipes. When adding plastic pipes to a metal plumbing system, make sure the electrical ground circuit is not broken. Use ground clamps and jumper wires, available at any hardware store, to bypass the plastic transition and complete the electrical ground circuit. Clamps must be firmly attached to bare metal on both sides of the plastic pipe.

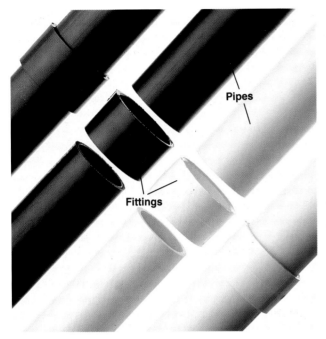

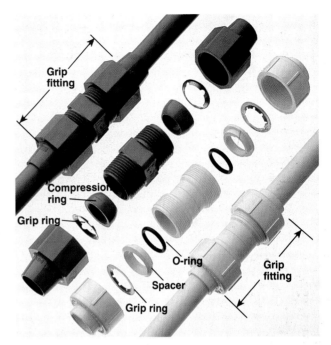

Solvent-glued fittings are used on rigid plastic pipes. Solvent dissolves a thin layer of plastic, and bonds the pipe and fitting together.

Grip fittings are used to join flexible PB pipes, and can also be used for CPVC pipes. Grip fittings come in two styles. One type (left) resembles a copper compression fitting. It has a metal grip ring and a plastic compression ring. The other type (right) has a rubber O-ring instead of a compression ring.

Plastic Pipe Grade Stamps

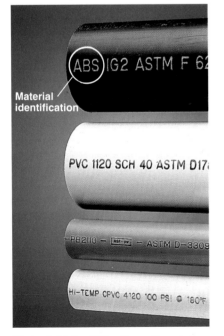

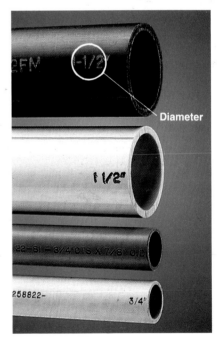

Material identification: For sink traps and drain pipes, use PVC or ABS pipe. For water supply pipes, use PB or CPVC pipe.

NSF rating: For sink traps and drains, choose PVC or ABS pipe that has a DWV (drain-waste-vent) rating from the National Sanitation Foundation (NSF). For water supply pipes, choose PB or CPVC pipe that has a PW (pressurized water) rating.

Pipe diameter: PVC and ABS pipes for drains usually have an inside diameter of 1¼'' to 4''. PB and CPVC pipes for water supply usually have an inside diameter of ½'' or ¾''.

Cutting & Fitting Plastic Pipe

Cut rigid ABS, PVC, or CPVC plastic pipes with a tubing cutter, or with any saw. Cuts must be straight to ensure watertight joints.

Rigid plastics are joined with plastic fittings and solvent glue. Use a solvent glue that is made for the type of plastic pipe you are installing. For example, do not use ABS solvent on PVC pipe. Some solvent glues, called "all-purpose" or "universal" solvents, may be used on all types of plastic pipe.

Solvent glue hardens in about 30 seconds, so test-fit all plastic pipes and fittings before gluing the first joint. For best results, and to meet Code requirements, the surfaces of plastic pipes and fittings should be dulled with liquid primer before they are joined.

Liquid solvent glues and primers are toxic and flammable. Provide adequate ventilation when fitting plastics, and store the products away from any source of heat.

Cut flexible PB pipes with a plastic tubing cutter, or with a knife. Make sure cut ends of pipe are straight. Join PB plastic pipes with plastic grip fittings. Grip fittings also are used to join rigid or flexible plastic pipes to copper plumbing pipes (page 15).

Everything You Need:

Tools: tape measure, felt-tipped pen, tubing cutter (or miter box or hacksaw), utility knife, channel-type pliers.

Materials: plastic pipe, fittings, emery cloth, plastic pipe primer, solvent glue, rag, petroleum jelly.

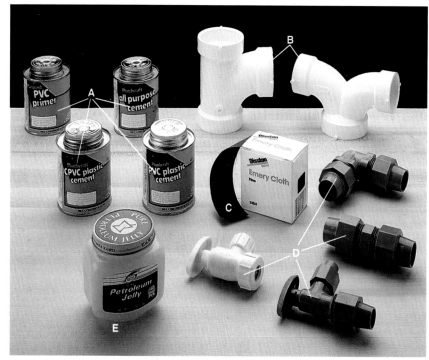

Specialty materials for plastics include: solvent glues and primer (A), solvent-glue fittings (B), emery cloth (C), plastic grip fittings (D), and petroleum jelly (E).

Measuring Plastic Pipe

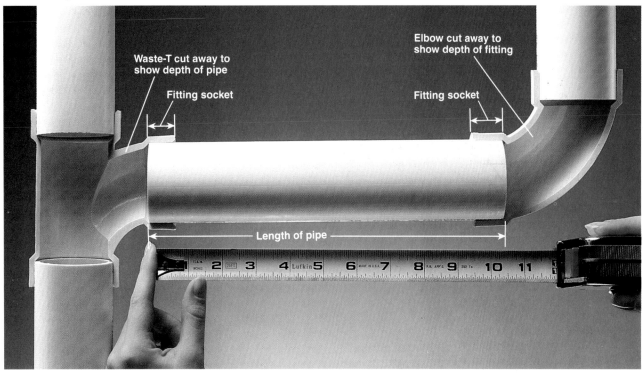

Waste-T cut away to show depth of pipe

Fitting socket

Elbow cut away to show depth of fitting

Fitting socket

Length of pipe

Find length of plastic pipe needed by measuring between the bottoms of the fitting sockets (fittings shown in cutaway). Mark the length on the pipe with a felt-tipped pen.

How to Cut Rigid Plastic Pipe

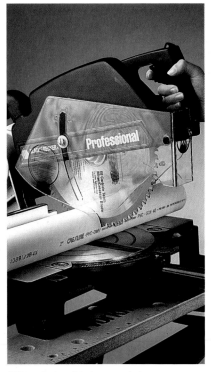

Tubing cutter: Tighten tool around pipe so cutting wheel is on marked line (page 19). Rotate tool around pipe, tightening screw every two rotations, until pipe snaps.

Miter box: Make straight cuts on all types of plastic pipe with a power or hand miter box.

Hacksaw: Clamp plastic pipe in a portable gripping bench or a vise, and keep the hacksaw blade straight while sawing.

How to Solvent-glue Rigid Plastic Pipe

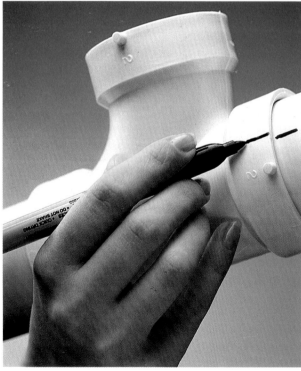

1 Remove rough burrs on cut ends of plastic pipe, using a utility knife.

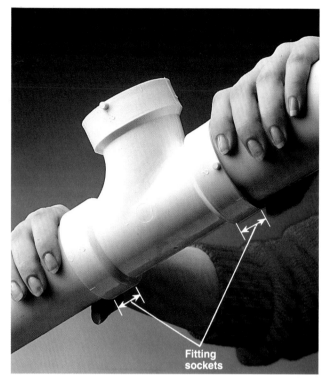

2 Test-fit all pipes and fittings. Pipes should fit tightly against the bottom of the fitting sockets.

3 Make alignment marks across each joint with a felt-tipped pen.

4 Mark depth of the fitting sockets on pipes. Take pipes apart.

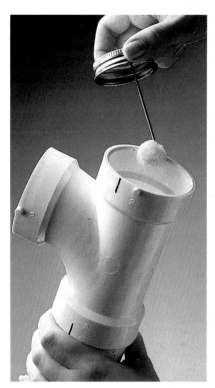

5 Clean ends of pipes with emery cloth if they are dirty or nicked.

6 Apply plastic pipe primer to the ends of the pipes. Primer dulls glossy surfaces to ensure a good seal, and is required by Code.

7 Apply plastic pipe primer to the insides of the fitting sockets.

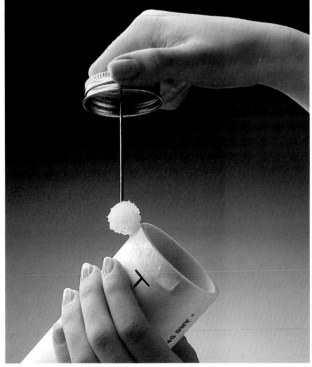

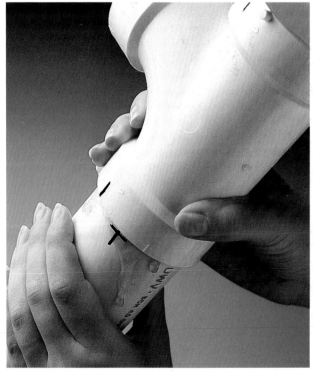

8 Solvent-glue each joint by applying a thick coat of solvent glue to end of pipe. Apply a thin coat of solvent glue to inside surface of fitting socket. Work quickly: solvent glue hardens in about 30 seconds.

9 Quickly position pipe and fitting so that alignment marks are offset by about 2 inches. Force pipe into fitting until the end fits flush against the bottom of the socket. Twist pipe into alignment (step 10).

(continued next page)

How to Solvent-glue Rigid Plastic Pipe (continued)

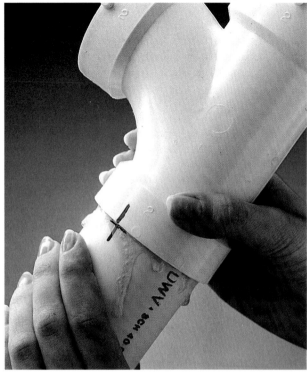

10 Spread solvent by twisting the pipe until marks are aligned. Hold pipe in place for about 20 seconds to prevent joint from slipping.

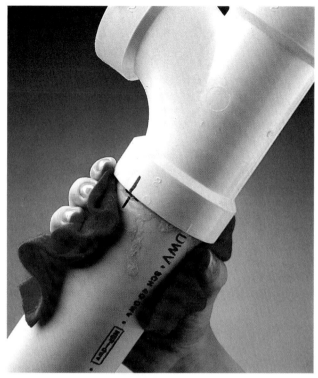

11 Wipe away excess solvent glue with a rag. Do not disturb joint for 30 minutes after gluing.

How to Cut & Fit Flexible Plastic Pipe

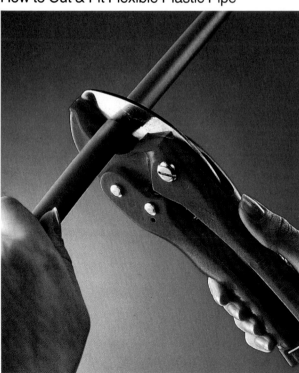

1 Cut flexible PB pipe with a plastic tubing cutter, available at home centers. (Flexible pipe also can be cut with a miter box or a sharp knife.) Remove any rough burrs with a utility knife.

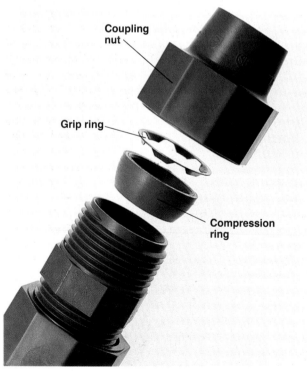

Coupling nut

Grip ring

Compression ring

2 Take each grip fitting apart and make sure that the grip ring and the compression ring or O-ring are positioned properly (page 27). Loosely reassemble the fitting.

3 Make a mark on the pipe showing the depth of the fitting socket, using a felt-tipped pen. Round off the edges of the pipe with emery cloth.

4 Lubricate the end of the pipe with petroleum jelly. Lubricated tip makes it easier to insert pipes into grip fittings.

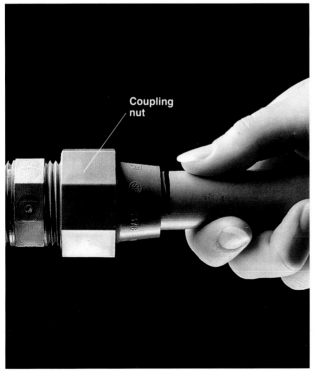

Coupling nut

5 Force end of pipe into fitting up to the mark on the pipe. Hand-tighten coupling nut.

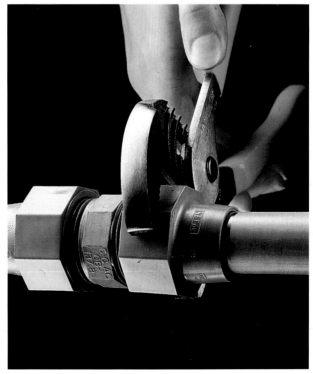

6 Tighten coupling nut about ½ turn with channel-type pliers. Turn on water and test the fitting. If the fitting leaks, tighten coupling nut slightly.

Working with Galvanized Iron

Galvanized iron pipe often is found in older homes, where it is used for water supply and small drain lines. It can be identified by the zinc coating that gives it a silver color, and by the threaded fittings used to connect pipes.

Galvanized iron pipes and fittings will corrode with age and eventually must be replaced. Low water pressure may be a sign that the insides of galvanized pipes have a buildup of rust. Blockage usually occurs in elbow fittings. Never try to clean the insides of galvanized iron pipes. Instead, remove and replace them as soon as possible.

Galvanized iron pipe and fittings are available at hardware stores and home improvement centers. Always specify the interior diameter (I.D.) when purchasing galvanized pipes and fittings. Pre-threaded pipes, called *nipples,* are available in lengths from 1" to 1 foot. If you need a longer length, have the store cut and thread the pipe to your dimensions.

Old galvanized iron can be difficult to repair. Fittings often are rusted in place, and what seems like a small job may become a large project. For example, cutting apart a section of pipe to replace a leaky fitting may reveal that adjacent pipes are also in need of replacement. If your job takes an unexpected amount of time, you can cap off any open lines and restore water to the rest of your house. Before you begin a repair, have on hand nipples and end caps that match your pipes.

Taking apart a system of galvanized iron pipes and fittings is time-consuming. Disassembly must start at the end of a pipe run, and each piece must be unscrewed before the next piece can be removed. Reaching the middle of a run to replace a section of pipe can be a long and tedious job. Instead, use a special three-piece fitting called a union. A union makes it possible to remove a section of pipe or a fitting without having to take the entire system apart.

Note: Galvanized iron is sometimes confused with "black iron." Both types have similar sizes and fittings. Black iron is used only for gas lines.

Everything You Need:

Tools: tape measure, reciprocating saw with metal-cutting blade or a hacksaw, pipe wrenches, propane torch, wire brush.

Materials: nipples, end caps, union fitting, pipe joint compound, replacement fittings (if needed).

Measure old pipe. Include ½" at each end for the threaded portion of the pipe inside fitting. Bring overall measurement to the store when shopping for replacement parts.

How to Remove & Replace a Galvanized Iron Pipe

1 Cut through galvanized iron pipe with a reciprocating saw and a metal-cutting blade, or with a hacksaw.

2 Hold fitting with one pipe wrench, and use another wrench to remove old pipe. Jaws of wrenches should face opposite directions. Always move wrench handle toward jaw opening.

3 Remove any corroded fittings using two pipe wrenches. With jaws facing in opposite directions, use one wrench to turn fitting and the other to hold the pipe. Clean pipe threads with a wire brush.

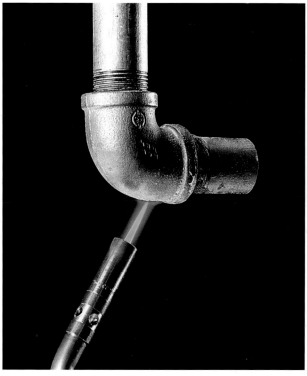

4 Heat stubborn fittings with a propane torch to make them easier to remove. Apply flame for 5 to 10 seconds. Protect wood or other flammable materials from heat, using a double layer of sheet metal (page 18).

(continued next page)

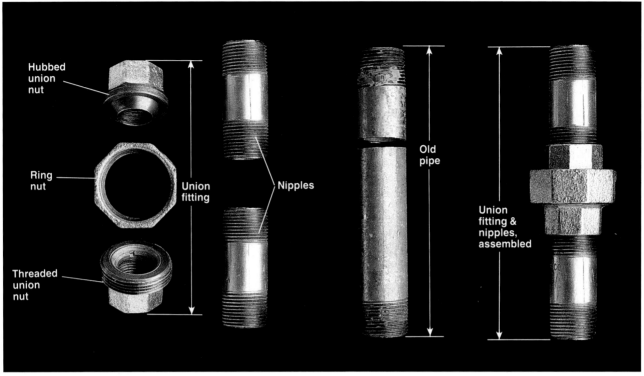

5 Replace a section of galvanized iron pipe with a union fitting and two threaded pipes (nipples). When assembled, the union and nipples must equal the length of the pipe that is being replaced.

6 Apply a bead of pipe joint compound around threaded ends of all pipes and nipples. Spread compound evenly over threads with fingertip.

7 Screw new fittings onto pipe threads. Tighten fittings with two pipe wrenches, leaving them about ⅛ turn out of alignment to allow assembly of union.

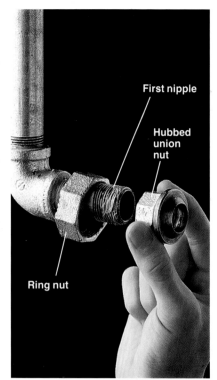

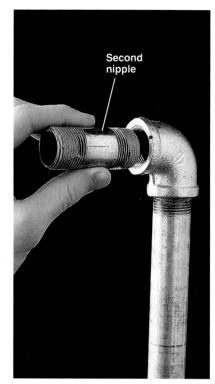

8 Screw first nipple into fitting, and tighten with pipe wrench.

9 Slide ring nut onto the installed nipple, then screw the hubbed union nut onto the nipple and tighten with a pipe wrench.

10 Screw second nipple onto other fitting. Tighten with pipe wrench.

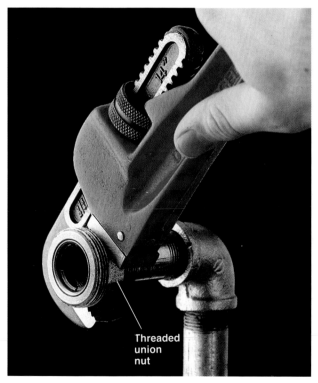

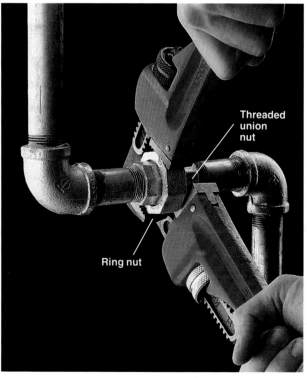

11 Screw threaded union nut onto second nipple. Tighten with a pipe wrench. Turn pipes into alignment, so that lip of hubbed union nut fits inside threaded union nut.

12 Complete the connection by screwing the ring nut onto the threaded union nut. Tighten ring nut with pipe wrenches.

Fixing Leaky Faucets

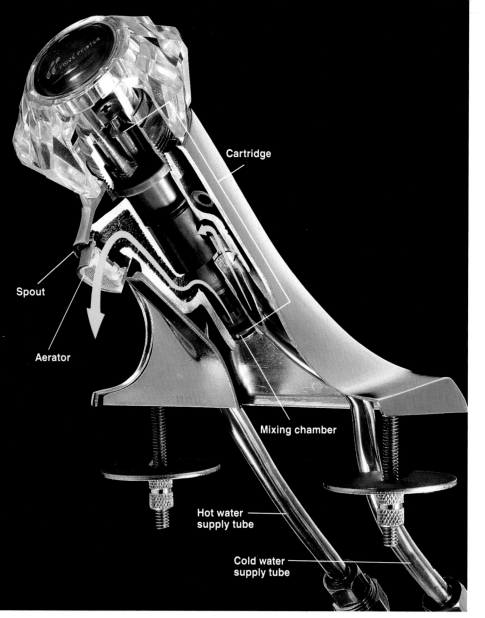

Typical faucet has a single handle attached to a hollow cartridge. The cartridge controls hot and cold water flowing from the supply tubes into the mixing chamber. Water is forced out the spout and through the aerator. When repairs are needed, replace the entire cartridge.

(labels on image) Cartridge / Spout / Aerator / Mixing chamber / Hot water supply tube / Cold water supply tube

A leaky faucet is the most common home plumbing problem. Leaks occur when washers, O-rings, or seals inside the faucet are dirty or worn. Fixing leaks is easy, but the techniques for making repairs will vary, depending on the design of the faucet. Before beginning work, you must first identify your faucet design and determine what replacement parts are needed.

There are four basic faucet designs: ball-type, cartridge, disc, or compression. Many faucets can be identified easily by outer appearance, but others must be taken apart before the design can be recognized.

The compression design is used in many double-handle faucets. Compression faucets all have washers or seals that must be replaced from time to time. These repairs are easy to make, and replacement parts are inexpensive.

Ball-type, cartridge, and disc faucets are all known as washerless faucets. Many washerless faucets are controlled with a single handle, although some cartridge models use two handles. Washerless faucets are more trouble-free than compression faucets, and are designed for quick repair.

When installing new faucet parts, make sure the replacements match the original parts. Replacement parts for popular washerless faucets are identified by brand name and model number. To ensure a correct selection, you may want to bring the worn parts to the store for comparison.

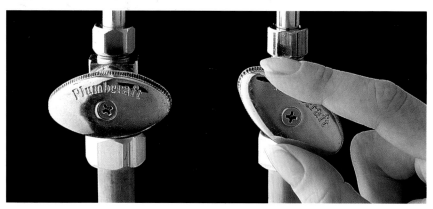

Turn off water before starting any faucet repair, using shutoff valves underneath faucet, or main service valve found near water meter (page 4). When opening shutoff valves after finishing repairs, keep faucet handle in open position to release trapped air. When water runs steadily, close faucet.

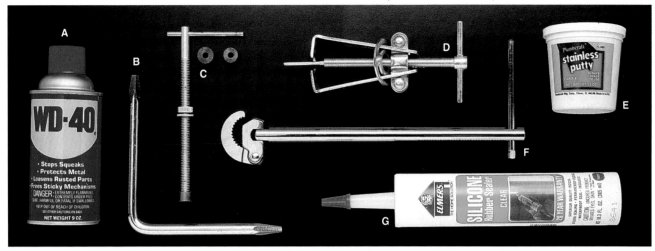

Specialty tools and materials for faucet repairs include: penetrating oil (A), seat wrench (B), seat-dressing (reamer) tool (C), handle puller (D), plumber's putty (E), basin wrench (F), silicone caulk (G).

How to Identify Faucet Designs

Ball-type faucet has a single handle over a dome-shaped cap. If your single-handle faucet is made by Delta or Peerless, it is probably a ball-type faucet. See pages 40 to 41 to fix a ball-type faucet.

Cartridge faucets are available in single-handle or double-handle models. Popular cartridge faucet brands include Price Pfister, Moen, Valley, and Aqualine. See pages 42 to 43 to fix a cartridge faucet.

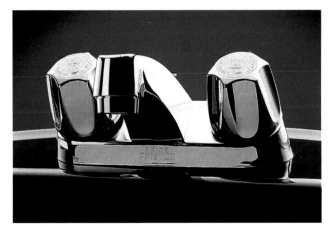

Compression faucet has two handles. When shutting the faucet off, you usually can feel a rubber washer being squeezed inside the faucet. Compression faucets are sold under many brand names. See pages 44 to 47 to fix a compression faucet.

Disc faucet has a single handle and a solid, chromed-brass body. If your faucet is made by American Standard or Reliant, it may be a disc faucet. See pages 48 to 49 to fix a disc faucet.

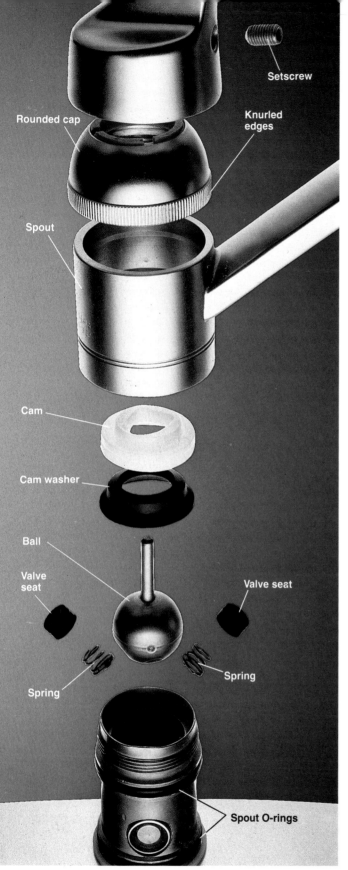

Fixing Ball-type Faucets

A ball-type faucet has a single handle, and is identified by the hollow metal or plastic ball inside the faucet body. Many ball-type faucets have a rounded cap with knurled edges located under the handle. If your faucet leaks from the spout and has this type of cap, first try tightening the cap with channel-type pliers. If tightening does not fix the leak, disassemble the faucet and install replacement parts.

Faucet manufacturers offer several types of replacement kits for ball-type faucets. Some kits contain only the springs and neoprene valve seats, while better kits also include the cam and cam washer.

Replace the rotating ball only if it is obviously worn or scratched. Replacement balls are either metal or plastic. Metal balls are slightly more expensive than plastic, but are more durable.

Remember to turn off the water before beginning work (page 38).

Everything You Need:

Tools: channel-type pliers, allen wrench, screwdriver, utility knife.

Materials: ball-type faucet repair kit, new rotating ball (if needed), masking tape, O-rings, heatproof grease.

Ball-type faucet has a hollow ball that controls the temperature and flow of water. Dripping at the faucet spout is caused by worn-out valve seats, springs, or a damaged ball. Leaks around the base of the faucet are caused by worn O-rings.

Repair kit for a ball-type faucet includes rubber valve seats, springs, cam, cam washer, and spout O-rings. Kit may also include small allen wrench tool used to remove faucet handle. Make sure kit is made for your faucet model. Replacement ball can be purchased separately, but is not needed unless old ball is obviously worn.

How to Fix a Ball-type Faucet

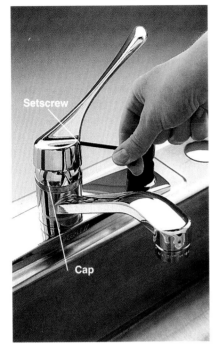

1 Loosen handle setscrew with an allen wrench. Remove handle to expose faucet cap.

2 Remove the cap with channel-type pliers. To prevent scratches to the shiny chromed finish, wrap masking tape around the jaws of the pliers.

3 Lift out the faucet cam, cam washer, and the rotating ball. Check the ball for signs of wear.

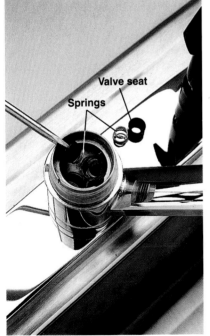

4 Reach into the faucet with a screwdriver and remove the old springs and neoprene valve seats.

5 Remove spout by twisting it upward, then cut off old O-rings. Coat new O-rings with heatproof grease, and install. Reattach spout, pressing downward until the collar rests on plastic slip ring. Install new springs and valve seats.

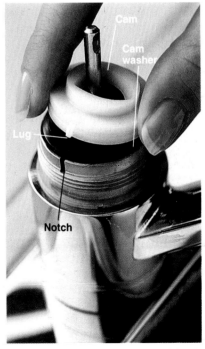

6 Insert ball, new cam washer, and cam. Small lug on cam should fit into notch on faucet body. Screw cap onto faucet and attach handle.

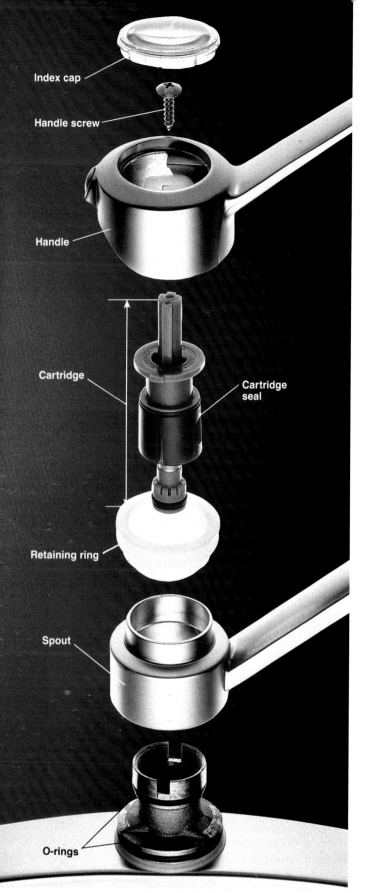

Index cap

Handle screw

Handle

Cartridge

Cartridge seal

Retaining ring

Spout

O-rings

Cartridge faucet has a hollow cartridge insert that lifts and rotates to control the flow and temperature of water. Dripping at the spout occurs when the cartridge seals become worn. Leaks around the base of the faucet are caused by worn O-rings.

Fixing Cartridge Faucets

A cartridge faucet is identified by the narrow metal or plastic cartridge inside the faucet body. Many single-handle faucets and some double-handle models use cartridge designs.

Replacing a cartridge is an easy repair that will fix most faucet leaks. Faucet cartridges come in many styles, so you may want to bring the old cartridge along for comparison when shopping for a replacement.

Make sure to insert the new cartridge so it is aligned in the same way as the old cartridge. If the hot and cold water controls are reversed, take the faucet apart and rotate the cartridge 180°.

Remember to turn off the water before beginning work (page 38).

Everything You Need:

Tools: screwdriver, channel-type pliers, utility knife.

Materials: replacement cartridge, O-rings, heat-proof grease.

Replacement cartridges come in dozens of styles. Cartridges are available for popular faucet brands, including (from left): Price-Pfister, Moen, Kohler. O-ring kits may be sold separately.

How to Fix a Cartridge Faucet

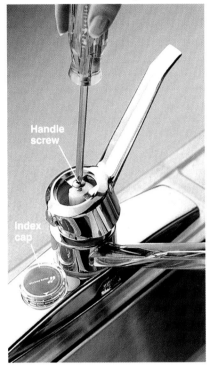

1 Pry off the index cap on top of faucet, and remove the handle screw underneath the cap.

2 Remove faucet handle by lifting it up and tilting it backwards.

3 Remove the threaded retaining ring with channel-type pliers. Remove any retaining clip holding cartridge in place.

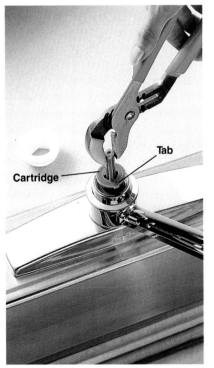

4 Grip top of the cartridge with channel-type pliers. Pull straight up to remove cartridge. Install replacement cartridge so that tab on cartridge faces forward.

5 Remove the spout by pulling up and twisting, then cut off old O-rings with a utility knife. Coat new O-rings with heatproof grease, and install.

6 Reattach the spout. Screw the retaining ring onto the faucet, and tighten with channel-type pliers. Attach the handle, handle screw, and index cap.

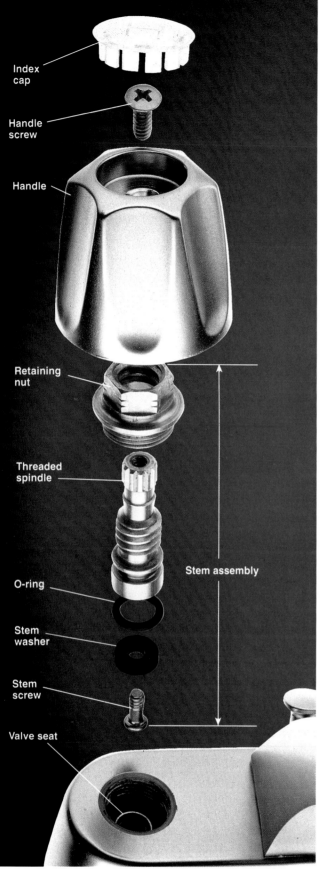

Index cap

Handle screw

Handle

Retaining nut

Threaded spindle

Stem assembly

O-ring

Stem washer

Stem screw

Valve seat

Fixing Compression Faucets

Compression faucets have separate controls for hot and cold water, and are identified by the threaded stem assemblies inside the faucet body. Compression stems come in many different styles, but all have some type of neoprene washer or seal to control water flow. Compression faucets leak when stem washers and seals become worn.

Older compression faucets often have corroded handles that are difficult to remove. A specialty tool called a handle puller makes this job easier. Handle pullers may be available at rental centers.

When replacing washers, also check the condition of the metal valve seats inside the faucet body. If the valve seats feel rough, they should be replaced or resurfaced.

Remember to turn off the water before beginning work (page 38).

Everything You Need:

Tools: screwdriver, handle puller (if needed), channel-type pliers, utility knife, seat wrench or seat-dressing tool (if needed).

Materials: universal washer kit, packing string, heatproof grease, replacement valve seats (if needed).

A compression faucet has a stem assembly that includes a retaining nut, threaded spindle, O-ring, stem washer, and stem screw. Dripping at the spout occurs when the washer becomes worn. Leaks around the handle are caused by a worn O-ring.

Universal washer kit contains parts needed to fix most types of compression faucets. Choose a kit that has an assortment of neoprene washers, O-rings, packing washers, and brass stem screws.

Tips for Fixing a Compression Faucet

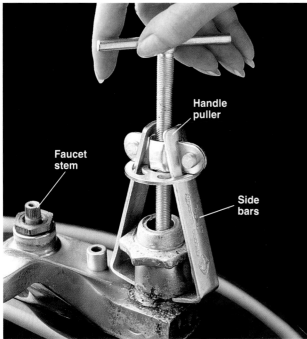

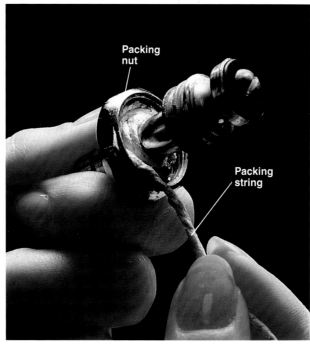

Remove stubborn handles with a handle puller. Remove the faucet index cap and handle screw, and clamp the side bars of the puller under the handle. Thread the puller into the faucet stem, and tighten until the handle comes free.

Packing string is used instead of an O-ring on some faucets. To fix leaks around the faucet handle, wrap new packing string around the stem, just underneath the packing nut or retaining nut.

Three Common Types of Compression Stems

Standard stem has a brass stem screw that holds either a flat or beveled neoprene washer to the end of the spindle. If stem screw is worn, it should be replaced.

Tophat stem has a snap-on neoprene diaphragm instead of a standard washer. Fix leaks by replacing the diaphragm.

Reverse-pressure stem has a beveled washer at the end of the spindle. To replace washer, unscrew spindle from rest of the stem assembly. Some stems have a small nut that holds washer.

How to Fix a Compression Faucet

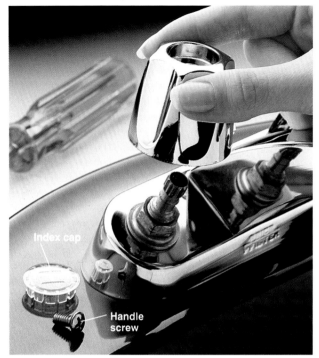

1 Remove index cap from top of faucet handle, and remove handle screw. Remove handle by pulling straight up. If necessary, use a handle puller to remove handle (page 45).

2 Unscrew the stem assembly from body of faucet, using channel-type pliers. Inspect valve seat for wear, and replace or resurface as needed (page opposite). If faucet body or stems are badly worn, it usually is best to replace the faucet (pages 50 to 53).

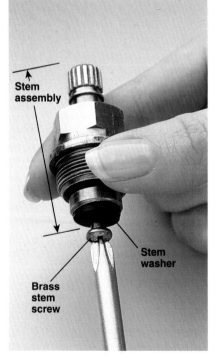

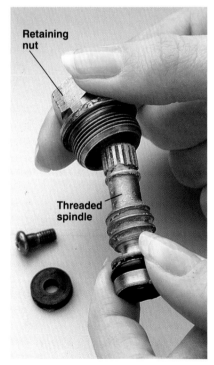

3 Remove the brass stem screw from the stem assembly. Remove worn stem washer.

4 Unscrew the threaded spindle from the retaining nut.

5 Cut off O-ring and replace with an exact duplicate. Install new washer and stem screw. Coat all parts with heatproof grease, then reassemble the faucet.

46

How to Replace Worn Valve Seats

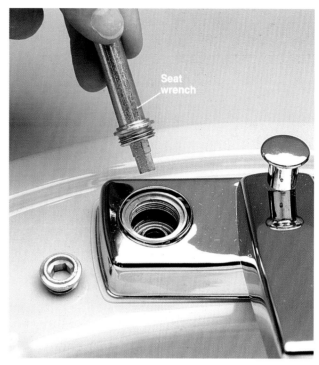

1 Check valve seat for damage by running a finger-tip around the rim of the seat. If the valve seat feels rough, replace the seat, or resurface it with a seat-dressing (reamer) tool (below).

2 Remove valve seat, using a seat wrench. Select end of wrench that fits seat, and insert into faucet. Turn counterclockwise to remove seat, then install an exact duplicate. If seat cannot be removed, resurface with a seat-dressing tool (below).

How to Resurface Valve Seats

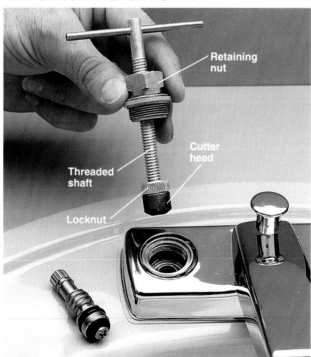

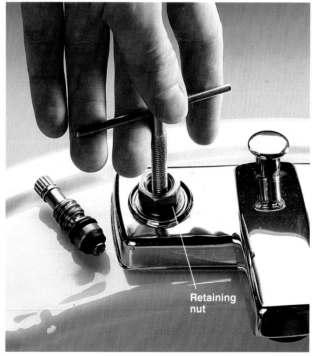

1 Select cutter head to fit the inside diameter of retaining nut. Slide retaining nut over threaded shaft of seat-dressing tool, then attach the locknut and cutter head to the shaft.

2 Screw retaining nut loosely into faucet body. Press the tool down lightly and turn tool handle clockwise two or three rotations. Reassemble faucet.

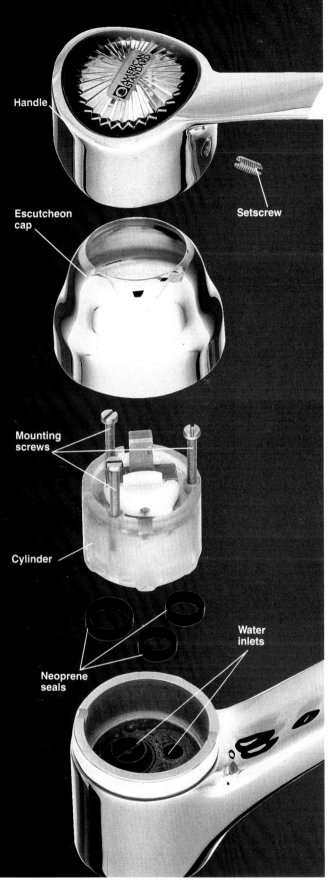

Fixing Disc Faucets

A disc faucet has a single handle and is identified by the wide cylinder inside the faucet body. The cylinder contains a pair of closely fitting ceramic discs that control the flow of water.

A ceramic disc faucet is a top-quality fixture that is easy to repair. Leaks usually can be fixed by lifting out the cylinder and cleaning the neoprene seals and the cylinder openings. Install a new cylinder only if the faucet continues to leak after cleaning.

After making repairs to a disc faucet, make sure handle is in the ON position, then open the shutoff valves slowly. Otherwise, ceramic discs can be cracked by the sudden release of air from the faucet. When water runs steadily, close the faucet.

Remember to turn off the water before beginning work (page 38).

Everything You Need:

Tools: screwdriver.

Materials: Scotch Brite® pad, replacement cylinder (if needed).

Disc faucet has a sealed cylinder containing two closely fitting ceramic discs. Faucet handle controls water by sliding the discs into alignment. Dripping at the spout occurs when the neoprene seals or cylinder openings are dirty.

Replacement cylinder for disc faucet is necessary only if faucet continues to leak after cleaning. Continuous leaking is caused by cracked or scratched ceramic discs. Replacement cylinders come with neoprene seals and mounting screws.

How to Fix a Ceramic Disc Faucet

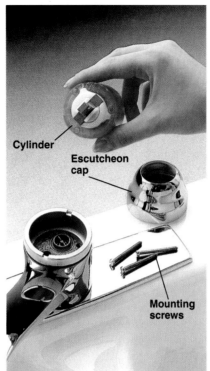

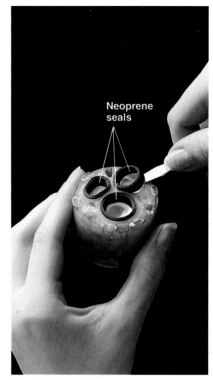

1 Rotate faucet spout to the side, and raise the handle. Remove the setscrew and lift off the handle.

2 Remove the escutcheon cap. Remove cartridge mounting screws, and lift out the cylinder.

3 Remove the neoprene seals from the cylinder openings.

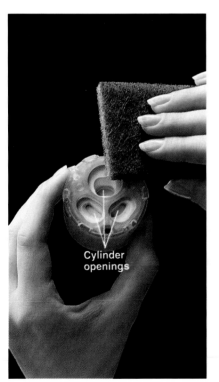

4 Clean the cylinder openings and the neoprene seals with a Scotch Brite® pad. Rinse cylinder with clear water.

5 Return seals to the cylinder openings, and reassemble faucet. Move handle to ON position, then slowly open shutoff valves. When water runs steadily, close faucet.

Install a new cylinder only if the faucet continues to leak after cleaning.

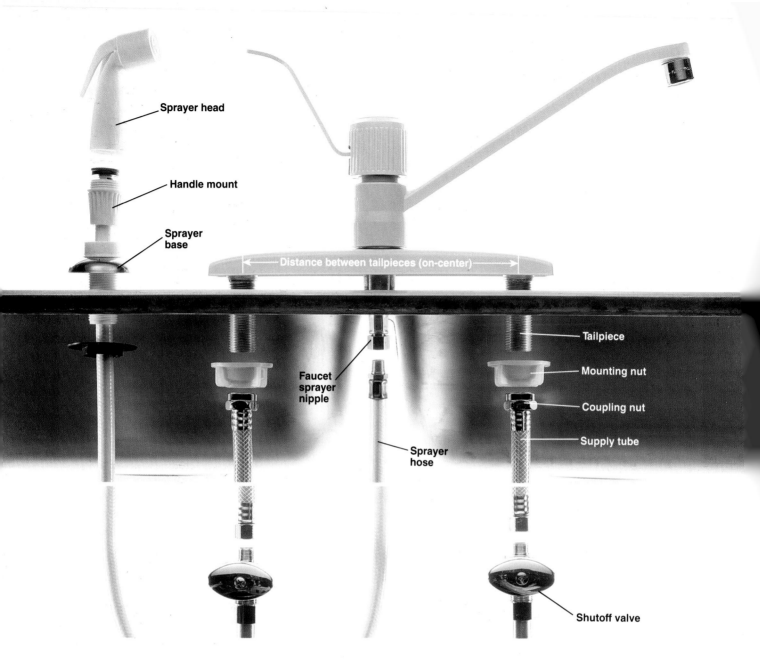

Sprayer head

Handle mount

Sprayer base

Distance between tailpieces (on-center)

Tailpiece

Mounting nut

Faucet sprayer nipple

Coupling nut

Supply tube

Sprayer hose

Shutoff valve

Replacing a Sink Faucet

Installing a new faucet is an easy project that takes about one hour. Before buying a new faucet, first find the diameter of the sink openings, and measure the distance between the tailpieces (measured on-center). Make sure the tailpieces of the new faucet match the sink openings.

When shopping for a new faucet, choose a model made by a reputable manufacturer. Replacement parts for a well-known brand will be easy to find if the faucet ever needs repairs. Better faucets have solid brass bodies. They are easy to install and provide years of trouble-free service. Some washerless models have lifetime warranties.

Always install new supply tubes when replacing a faucet. Old supply tubes should not be reused. If

water pipes underneath the sink do not have shut-off valves, you may choose to install the valves while replacing the faucet (pages 54 to 55).

Remember to turn off the water before beginning work (page 38).

Everything You Need:

Tools: basin wrench or channel-type pliers, putty knife, caulk gun, adjustable wrenches.

Materials: penetrating oil, silicone caulk or plumber's putty, two flexible supply tubes.

How to Remove an Old Sink Faucet

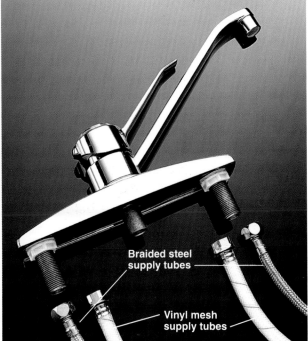

1 Spray penetrating oil on tailpiece mounting nuts and supply tube coupling nuts. Remove the coupling nuts with a basin wrench or channel-type pliers.

2 Remove the tailpiece mounting nuts with a basin wrench or channel-type pliers. Basin wrench has a long handle that makes it easy to work in tight areas.

3 Remove faucet. Use a putty knife to clean away old putty from surface of sink.

Faucet Hookup Variations

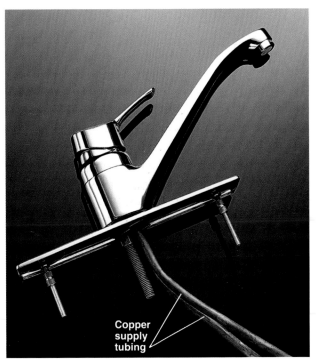

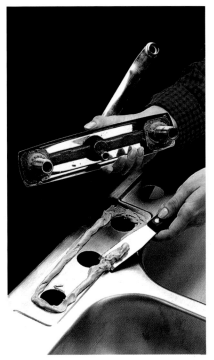

New faucet without supply tubes: Buy two supply tubes. Supply tubes are available in braided steel or vinyl mesh (shown above), PB plastic, or chromed copper (page 54).

New faucet with preattached copper supply tubing: Make water connections by attaching the supply tubing directly to the shutoff valves with compression fittings (page 53).

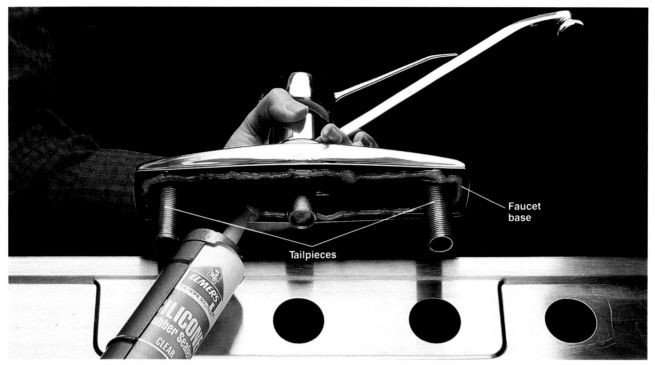

Faucet base

Tailpieces

1 Apply a ¼" bead of silicone caulk or plumber's putty around the base of the faucet. Insert the faucet tailpieces into the sink openings. Position the faucet so base is parallel to back of sink, and press the faucet down to make sure caulk forms a good seal.

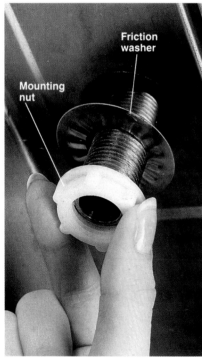

Friction washer

Mounting nut

Tailpiece

Coupling nut

Supply tube

Supply tube

Shutoff valve

2 Screw the metal friction washers and the mounting nuts onto the tailpieces, then tighten with a basin wrench or channel-type pliers. Wipe away excess caulk around base of faucet.

3 Connect flexible supply tubes to faucet tailpieces. Tighten coupling nuts with a basin wrench or channel-type pliers, but be careful not to overtighten.

4 Attach supply tubes to shutoff valves, using compression fittings (pages 24 to 25). Hand-tighten nuts, then use an adjustable wrench to tighten nuts ¼ turn. If necessary, hold valve with another wrench while tightening.

How to Connect a Faucet with Preattached Supply Tubing

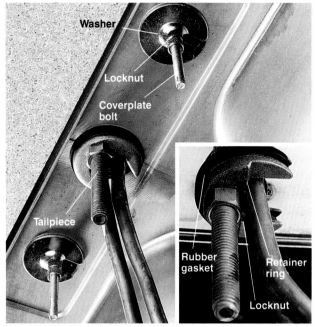

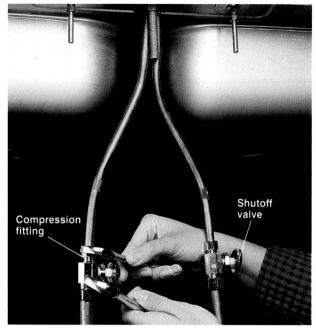

1 Attach faucet to sink by placing rubber gasket, retainer ring, and locknut onto threaded tailpiece. Tighten locknut with a basin wrench or channel-type pliers. Some center-mounted faucets have a decorative coverplate. Secure coverplate from underneath with washers and locknuts screwed onto coverplate bolts.

2 Connect preattached supply tubing to shutoff valves with compression fittings (pages 24 to 25). Red-coded tube should be attached to the hot water pipe, blue-coded tube to the cold water pipe.

How to Attach a Sink Sprayer

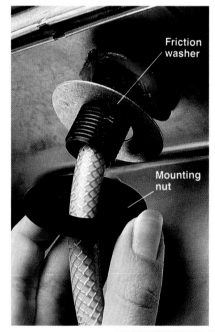

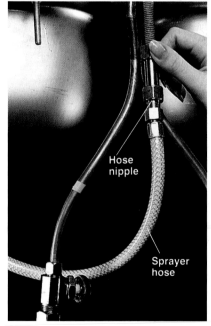

1 Apply a ¼" bead of plumber's putty or silicone caulk to bottom edge of sprayer base. Insert tailpiece of sprayer base into sink opening.

2 Place friction washer over tailpiece. Screw the mounting nut onto tailpiece and tighten with a basin wrench or channel-type pliers. Wipe away excess putty around base of sprayer.

3 Screw sprayer hose onto the hose nipple on the bottom of the faucet. Tighten ¼ turn, using a basin wrench or channel-type pliers.

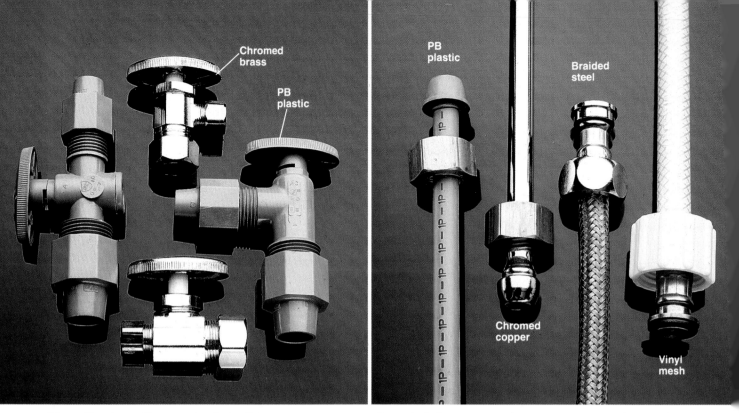

Shutoff valves allow you to shut off the water to an individual fixture so it can be repaired. They can be made from durable chromed brass or lightweight plastic. Shutoff valves come in ½" and ¾" diameters to match common water pipe sizes.

Supply tubes are used to connect water pipes to faucets, toilets, and other fixtures. They come in 12", 20", and 30" lengths. PB plastic and chromed copper tubes are inexpensive. Braided steel and vinyl mesh supply tubes are easy to install.

Installing Shutoff Valves & Supply Tubes

Worn-out shutoff valves or supply tubes can cause water to leak underneath a sink or other fixture. First, try tightening the fittings with an adjustable wrench. If this does not fix the leak, replace the shutoff valves and supply tubes.

Shutoff valves are available in several fitting types. For copper pipes, valves with compression-type fittings (pages 24 to 25) are easiest to install. For plastic pipes (pages 26 to 33), use grip-type valves. For galvanized iron pipes (pages 34 to 37), use valves with female threads.

Older plumbing systems often were installed without fixture shutoff valves. When repairing or replacing plumbing fixtures, you may want to install shutoff valves if they are not already present.

Everything You Need:

Tools: hacksaw, tubing cutter, adjustable wrench, tubing bender, felt-tipped pen.

Materials: shutoff valves, supply tubes, pipe joint compound.

How to Install Shutoff Valves & Supply Tubes

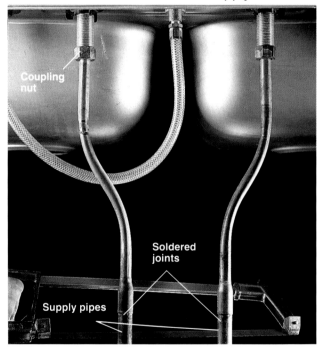

1 Turn off water at the main shutoff valve (page 4). Remove old supply pipes. If pipes are soldered copper, cut them off just below the soldered joint, using a hacksaw or tubing cutter. Make sure the cuts are straight. Unscrew the coupling nuts, and discard old pipes.

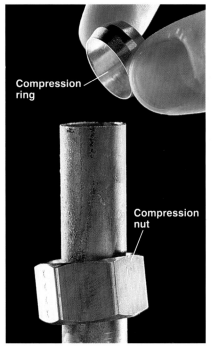

2 Slide a compression nut and compression ring over copper water pipe. Threads of nut should face end of pipe.

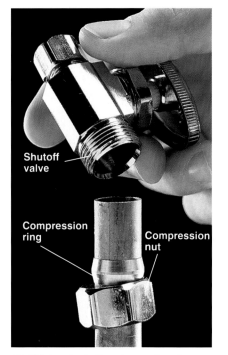

3 Slide shutoff valve onto pipe. Apply a layer of pipe joint compound to compression ring. Screw the compression nut onto the shutoff valve and tighten with an adjustable wrench.

4 Bend chromed copper supply tube to reach from the tailpiece of the fixture to the shutoff valve, using a tubing bender (page 17). Bend the tube slowly to avoid crimping the metal.

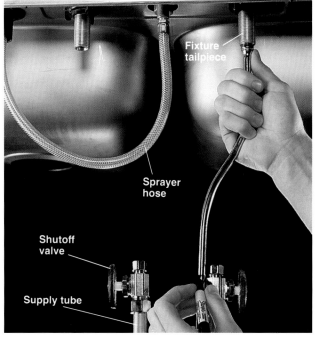

5 Position the supply tube between fixture tailpiece and shutoff valve, and mark tube to length. Cut supply tube with a tubing cutter (page 19).

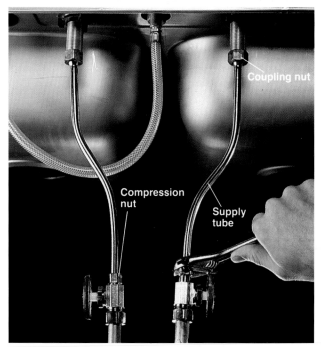

6 Attach bell-shaped end of supply tube to fixture tailpiece with coupling nut, then attach other end to shutoff valve with compression ring and nut (pages 24 to 25). Tighten all fittings with adjustable wrench.

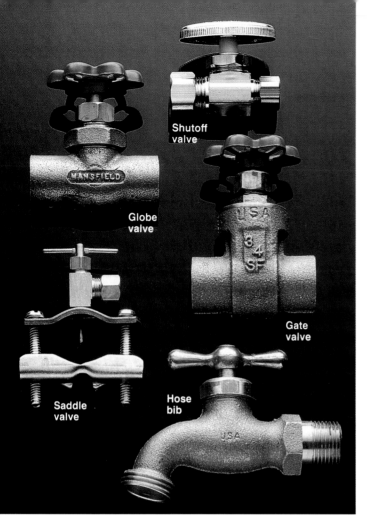

Shutoff valve

Globe valve

Gate valve

Saddle valve

Hose bib

Repairing Valves & Hose Bibs

Valves make it possible to shut off water at any point in the supply system. If a pipe breaks or a plumbing fixture begins to leak, you can shut off water to the damaged area so it can be repaired. A hose bib is a faucet with a threaded spout, often used to connect rubber utility or appliance hoses. This connection requires a vacuum breaker (page 15).

Valves and hose bibs leak when washers or seals wear out. Replacement parts can be found in the same universal washer kits used to repair compression faucets (page 44). Coat replacement washers with heatproof grease to keep them soft and prevent cracking.

Remember to turn off the water before beginning work (page 38).

Everything You Need:

Tools: screwdriver, adjustable wrench.

Materials: universal washer kit, heatproof grease.

How to Fix a Leaky Hose Bib

Packing nut

1 Remove the handle screw, and lift off the handle. Unscrew the packing nut with an adjustable wrench.

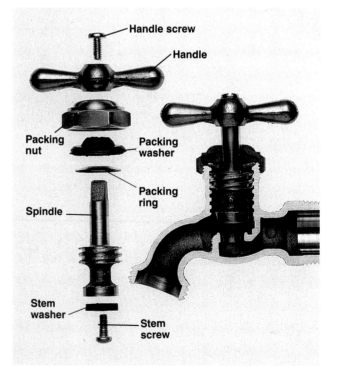

Handle screw

Handle

Packing nut

Packing washer

Packing ring

Spindle

Stem washer

Stem screw

2 Unscrew the spindle from the valve body. Remove the stem screw and replace the stem washer. Replace the packing washer, and reassemble the valve.

Common Types of Valves

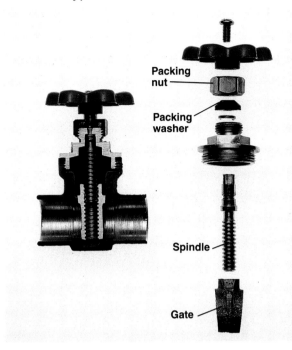

Gate valve has a movable brass wedge, or "gate," that screws up and down to control water flow. Gate valves may develop leaks around the handle. Repair leaks by replacing the packing washer or packing string found underneath the packing nut.

Globe valve has a curved chamber. Repair leaks around the handle by replacing the packing washer. If valve does not fully stop water flow when closed, replace the stem washer.

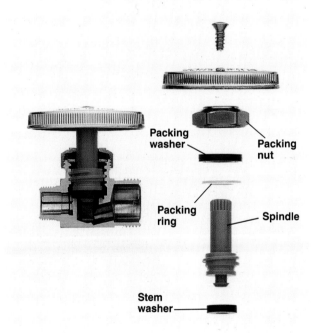

Shutoff valve controls water supply to a single fixture (pages 54 to 55). Shutoff valve has a plastic spindle with a packing washer and a snap-on stem washer. Repair leaks around the handle by replacing the packing washer. If valve does not fully stop water flow when closed, replace the stem washer.

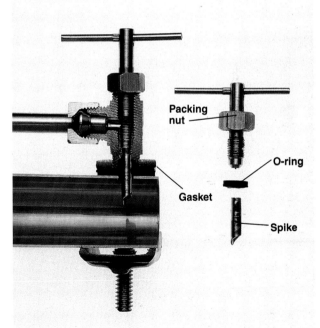

Saddle valve is a small fitting often used to connect a refrigerator icemaker or sink-mounted water filter to a copper water pipe. Saddle valve contains a hollow metal spike that punctures water pipe when valve is first closed. Fitting is sealed with a rubber gasket. Repair leaks around the handle by replacing the O-ring under the packing nut.

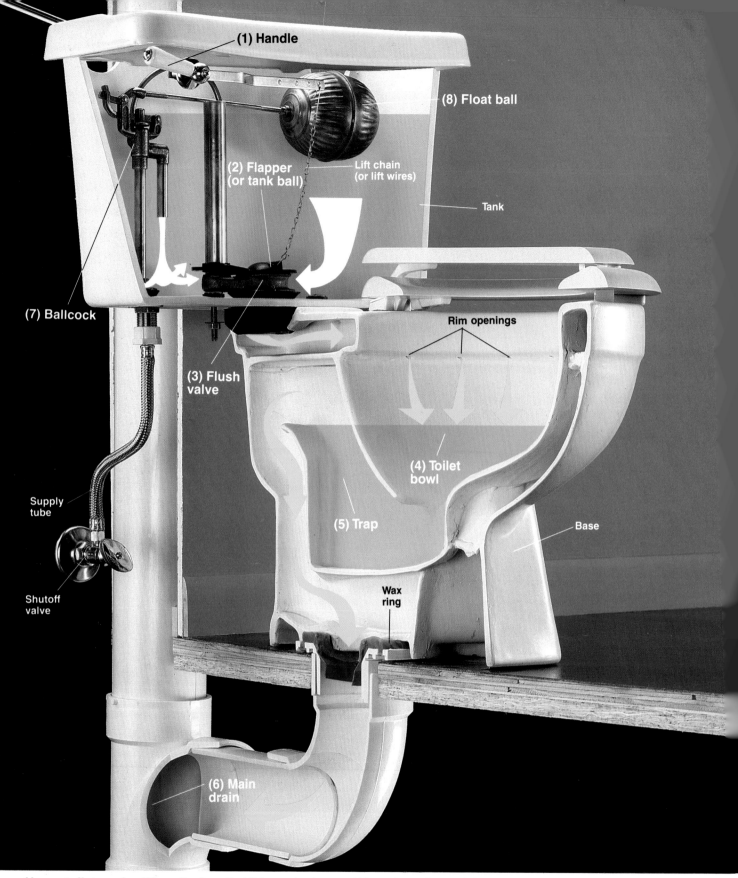

(1) Handle

(8) Float ball

(2) Flapper (or tank ball)

Lift chain (or lift wires)

Tank

(7) Ballcock

(3) Flush valve

Rim openings

Supply tube

(4) Toilet bowl

(5) Trap

Base

Shutoff valve

Wax ring

(6) Main drain

How a toilet works: When the **handle (1)** is pushed, the lift chain raises a rubber seal, called a **flapper** or **tank ball (2)**. Water in the tank rushes down through the **flush valve opening (3)** in the bottom of the tank, into the **toilet bowl (4)**. Waste water in the bowl is forced through the **trap (5)** into the **main drain (6)**.

When the toilet tank is empty, the flapper seals the tank, and a water supply valve, called a **ballcock (7)**, refills the toilet tank. The ballcock is controlled by a **float ball (8)** that rides on the surface of the water. When the tank is full, the float ball automatically shuts off the ballcock.

Fixing a Running Toilet

The sound of continuously running water occurs if fresh water continues to enter the toilet tank after the flush cycle is complete. A running toilet can waste 20 or more gallons of fresh water each day.

To fix a running toilet, first jiggle the toilet handle. If the sound of running water stops, then either the handle or the lift wires (or lift chain) need to be adjusted (page opposite).

If the sound of running water does not stop when the handle is jiggled, then remove the tank cover and check to see if the float ball is touching the side of the tank. If necessary, bend the float arm to reposition the float ball away from the side of the tank. Make sure the float ball is not leaking. To check for leaks, unscrew the float ball and shake it gently. If there is water inside the ball, replace it.

If these minor adjustments do not fix the problem, then you will need to adjust or repair the ballcock or the flush valve (photo, right). Follow the directions on the following pages.

Everything You Need:

Tools: screwdriver, small wire brush, sponge, adjustable wrenches, spud wrench or channel-type pliers.

Materials: universal washer kit, ballcock (if needed), ballcock seals, emery cloth, Scotch Brite® pad, flapper or tank ball, flush valve (if needed).

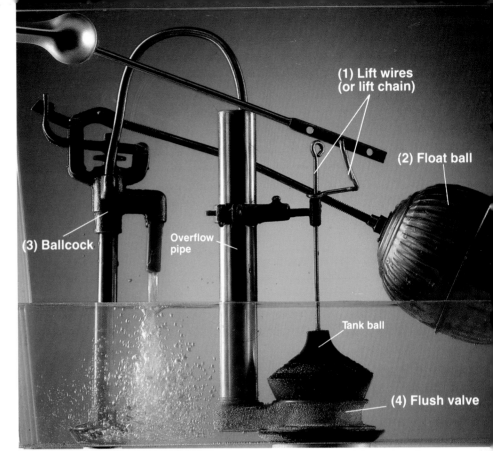

(1) Lift wires (or lift chain)
(2) Float ball
Overflow pipe
(3) Ballcock
Tank ball
(4) Flush valve

The sound of continuously running water can be caused by several different problems: if the **lift wire (1)** (or lift chain) is bent or kinked; if the **float ball (2)** leaks or rubs against the side of the tank; if a faulty **ballcock (3)** does not shut off the fresh water supply; or if the **flush valve (4)** allows water to leak down into the toilet bowl. First, check the lift wires and float ball. If making simple adjustments and repairs to these parts does not fix the problem, then you will need to repair the ballcock or flush valve (photo, below).

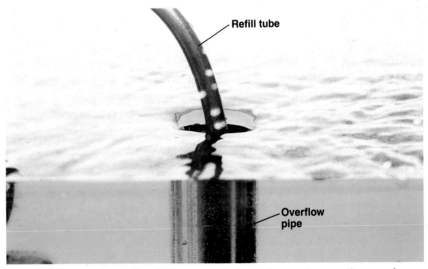

Refill tube
Overflow pipe

Check the overflow pipe if the sound of running water continues after the float ball and lift wires are adjusted. If you see **water flowing into the overflow pipe**, the ballcock needs to be repaired. First, adjust ballcock to lower the water level in the tank (page 60). If problem continues, repair or replace the ballcock (pages 61 to 62). If **water is not flowing into the overflow pipe**, then the flush valve needs to be repaired (page 63). First check the tank ball (or flapper) for wear, and replace if necessary. If problem continues, replace the flush valve.

How to Adjust a Ballcock to Set Water Level

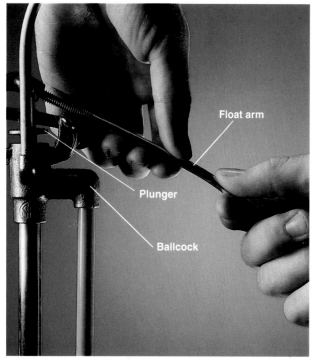

Traditional plunger-valve ballcock is made of brass. Water flow is controlled by a plunger attached to the float arm and ball. Lower the water level by bending the float arm downward slightly. Raise the water level by bending float arm upward.

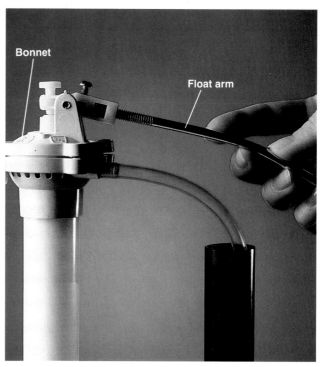

Diaphragm ballcock usually is made of plastic, and has a wide bonnet that contains a rubber diaphragm. Lower the water level by bending the float arm downward slightly. Raise the water level by bending float arm upward.

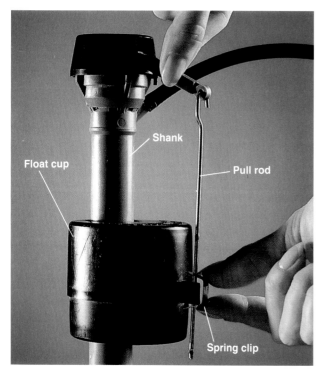

Float cup ballcock is made of plastic, and is easy to adjust. Lower the water level by pinching spring clip on pull rod, and moving float cup downward on the ballcock shank. Raise the water level by moving the cup upward.

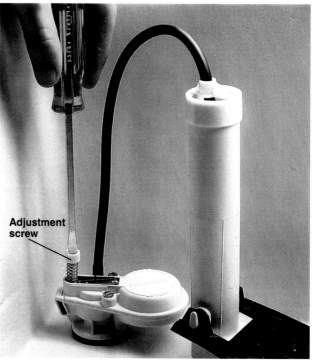

Floatless ballcock controls water level with a pressure-sensing device. Lower the water level by turning the adjustment screw counterclockwise, ½ turn at a time. Raise water level by turning screw clockwise. **NOTE:** This adjustment is a temporary repair. These submerged ballcocks do not meet Code and should be replaced.

How to Repair a Plunger-valve Ballcock

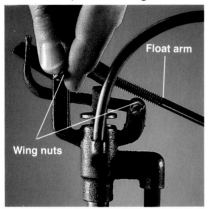

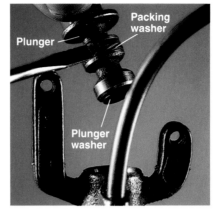

1 Shut off the water, and flush to empty the tank. Remove the wing nuts on the ballcock. Slip out the float arm.

2 Pull up on plunger to remove it. Pry out packing washer or O-ring. Pry out plunger washer. (Remove stem screw, if necessary.)

3 Install replacement washers. Clean sediment from inside of ballcock with a wire brush. Reassemble ballcock.

How to Repair a Diaphragm Ballcock

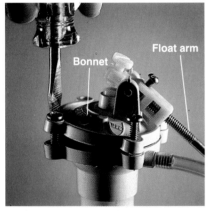

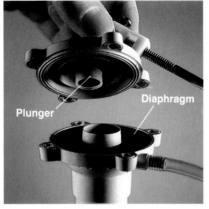

1 Shut off the water, and flush to empty the tank. Remove the screws from the bonnet.

2 Lift off float arm with bonnet attached. Check diaphragm and plunger for wear.

3 Replace any stiff or cracked parts. If assembly is badly worn, replace the entire ballcock (page 62).

How to Repair a Float Cup Ballcock

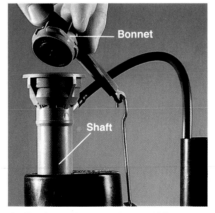

1 Shut off the water, and flush to empty the tank. Remove the ballcock cap.

2 Remove bonnet by pushing down on shaft and turning counterclockwise. Clean out sediment inside ballcock with wire brush.

3 Replace the seal. If assembly is badly worn, replace the entire ballcock (page 62).

How to Install a New Ballcock

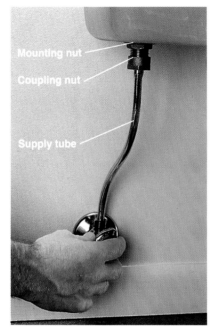

1 Shut off water, and flush toilet to empty tank. Use a sponge to remove remaining water. Disconnect supply tube coupling nut and ballcock mounting nut with adjustable wrench. Remove old ballcock.

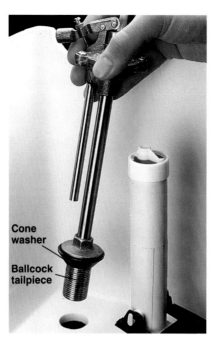

2 Attach cone washer to new ballcock tailpiece and insert tailpiece into tank opening.

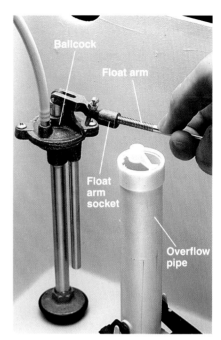

3 Align the float arm socket so that float arm will pass behind overflow pipe. Screw float arm onto ballcock. Screw float ball onto float arm.

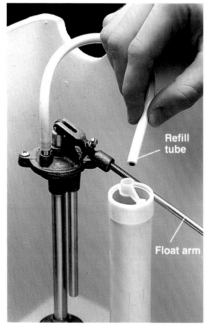

4 Bend or clip refill tube so tip is inside overflow pipe.

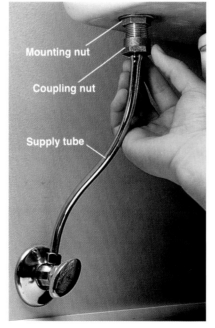

5 Screw mounting nut and supply tube coupling nut onto ballcock tailpiece, and tighten with an adjustable wrench. Turn on the water, and check for leaks.

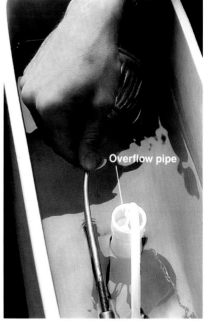

6 Adjust the water level in the tank so it is about ½" below top of the overflow pipe (page 60).

How to Adjust & Clean a Flush Valve

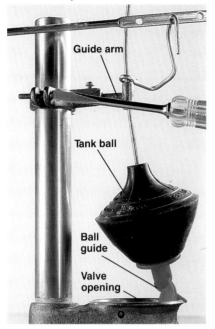

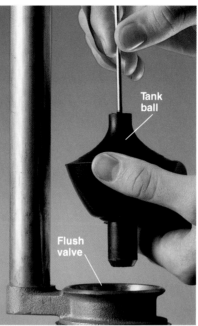

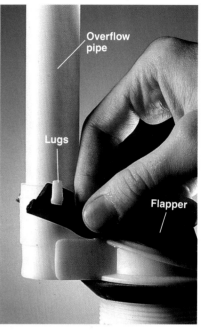

Adjust tank ball (or flapper) so it is directly over flush valve. Tank ball has a guide arm that can be loosened so that tank ball can be repositioned. (Some tank balls have a ball guide that helps seat the tank ball into the flush valve.)

Replace the tank ball if it is cracked or worn. Tank balls have a threaded fitting that screws onto the lift wire. Clean opening of the flush valve, using emery cloth (for brass valves) or a Scotch Brite® pad (for plastic valves).

Replace flapper if it is worn. Flappers are attached to small lugs on the sides of overflow pipe.

How to Install a New Flush Valve

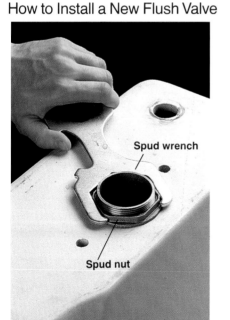

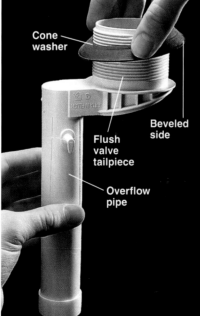

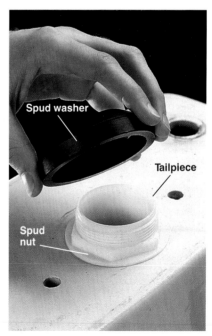

1 Shut off water, disconnect ballcock (page opposite, step 1), and remove toilet tank (page 65, steps 1 & 2). Remove old flush valve by unscrewing spud nut with spud wrench or channel-type pliers.

2 Slide cone washer onto tailpiece of new flush valve. Beveled side of cone washer should face end of tailpiece. Insert flush valve into tank opening so that overflow pipe faces ballcock.

3 Screw spud nut onto tailpiece of flush valve, and tighten with a spud wrench or channel-type pliers. Place soft spud washer over tailpiece, and reinstall toilet tank (pages 66 to 67).

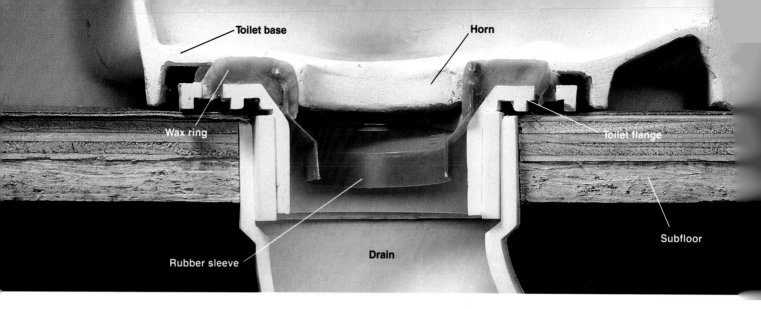

Toilet base · Horn · Wax ring · Toilet flange · Rubber sleeve · Drain · Subfloor

Fixing a Leaking Toilet

Water leaking onto the floor around a toilet may be caused by several different problems. The leaking must be fixed as soon as possible to prevent moisture from damaging the subfloor.

First, make sure all connections are tight. If moisture drips from the tank during humid weather, it is probably condensation. Fix this "sweating" problem by insulating the inside of the tank with foam panels. A crack in a toilet tank also can cause leaks. A cracked tank must be replaced.

Water seeping around the base of a toilet can be caused by an old wax ring that no longer seals against the drain (photo, above), or by a cracked toilet base. If leaking occurs during or just after a flush, replace the wax ring. If leaking is constant, the toilet base is cracked and must be replaced.

New toilets sometimes are sold with flush valves and ballcocks already installed. If these parts are not included, you will need to purchase them. When buying a new toilet, consider a water-saver design. Water-saver toilets use less than half the water needed by a standard toilet.

Everything You Need:

Tools: sponge, adjustable wrench, putty knife, ratchet wrench, screwdriver.

Materials: tank liner kit, abrasive cleanser, rag, wax ring, plumber's putty. *For new installation:* new toilet, toilet handle, ballcock, flush valve, tank bolts, toilet seat.

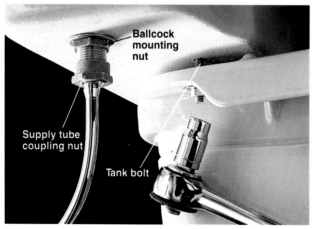

Ballcock mounting nut · Supply tube coupling nut · Tank bolt

Tighten all connections slightly. Tighten nuts on tank bolts with a ratchet wrench. Tighten ballcock mounting nut and supply tube coupling nut with an adjustable wrench. **Caution: overtightening tank bolts may crack the toilet tank.**

Insulate toilet tank to prevent "sweating," using a toilet liner kit. First, shut off water, drain tank, and clean inside of tank with abrasive cleanser. Cut plastic foam panels to fit bottom, sides, front, and back of tank. Attach panels to tank with adhesive (included in kit). Let adhesive cure as directed.

How to Remove & Replace a Wax Ring & Toilet

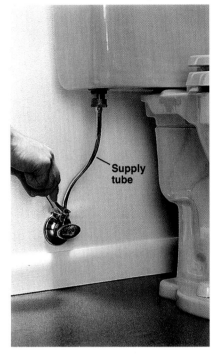

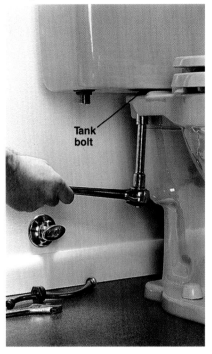

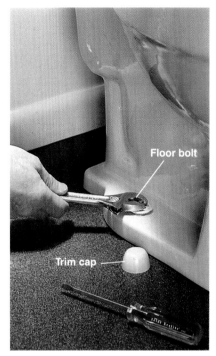

1 Turn off water, and flush to empty toilet tank. Use a sponge to remove remaining water in tank and bowl. Disconnect supply tube with an adjustable wrench.

2 Remove the nuts from the tank bolts with a ratchet wrench. Carefully remove the tank and set it aside.

3 Pry off the floor bolt trim caps at the base of the toilet. Remove the floor nuts with an adjustable wrench.

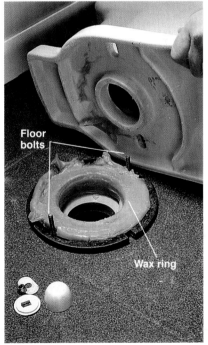

4 Straddle the toilet and rock the bowl from side to side until the seal breaks. Carefully lift the toilet off the floor bolts and set it on its side. Small amount of water may spill from the toilet trap.

5 Remove old wax from the toilet flange in the floor. Plug the drain opening with a damp rag to prevent sewer gases from rising into the house.

6 If old toilet will be reused, clean old wax and putty from the horn and the base of the toilet.

(continued next page)

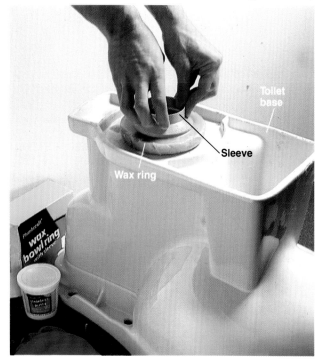

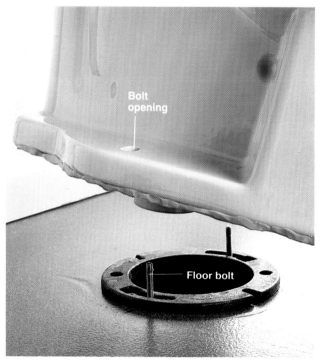

7 Turn stool upside down. Place new wax ring over drain horn. If ring has a rubber or plastic sleeve, sleeve should face away from toilet. Apply a bead of plumber's putty to bottom edge of toilet base.

8 Position the toilet over drain so that the floor bolts fit through the openings in the base of the toilet.

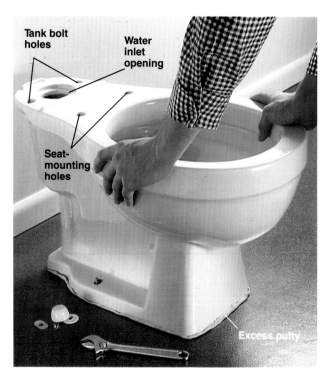

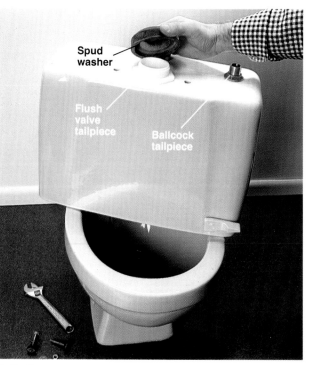

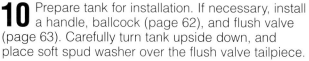

9 Press down on toilet base to compress wax and putty. Thread washers and nuts onto floor bolts, and tighten with adjustable wrench until snug. **Caution: overtightening nuts may crack the base.** Wipe away excess plumber's putty. Cover nuts with trim caps.

10 Prepare tank for installation. If necessary, install a handle, ballcock (page 62), and flush valve (page 63). Carefully turn tank upside down, and place soft spud washer over the flush valve tailpiece.

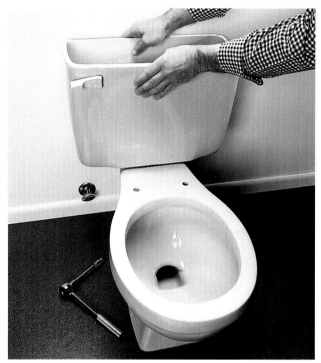

11 Turn tank right side up and position it on rear of toilet base so that spud washer is centered in water inlet opening.

12 Line up the tank bolt holes with holes in base of toilet. Slide rubber washers onto the tank bolts and place the bolts through holes. From underneath the tank, thread washers and nuts onto the bolts.

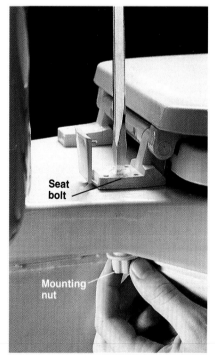

13 Tighten nuts with ratchet wrench until tank is snug. Use caution when tightening nuts: most toilet tanks rest on the spud washer, not directly on the toilet base.

14 Attach the water supply tube to the ballcock tailpiece with an adjustable wrench (page 62). Turn on the water and test toilet. Tighten tank bolts and water connections, if necessary.

15 Position the new toilet seat, if needed, inserting seat bolts into mounting holes in toilet. Screw mounting nuts onto the seat bolts, and tighten.

Tub & Shower Plumbing

Tub and shower faucets have the same basic designs as sink faucets, and the techniques for repairing leaks are the same as described in the faucet repair section of this book (pages 39 to 49). To identify your faucet design, you may need to take off the handle and disassemble the faucet.

When a tub and shower are combined, the shower head and the tub spout share the same hot and cold water supply lines and handles. Combination faucets are available as three-handle, two-handle,

Tub & Shower Combination Faucets

Shower head

Diverter valve

Hot water supply line

Cold water supply line

Tub spout

Three-handle faucet (pages 70 to 71) has valves that are either compression or cartridge design.

or single-handle types (below). The number of handles gives clues as to the design of the faucets and the kinds of repairs that may be necessary.

With combination faucets, a diverter valve or gate diverter is used to direct water flow to the tub spout or the shower head. On three-handle faucet types, the middle handle controls a diverter valve. If water does not shift easily from tub spout to shower head, or if water continues to run out the spout when the shower is on, the diverter valve probably needs to be cleaned and repaired (pages 70 to 71).

Two-handle and single-handle types use a gate diverter that is operated by a pull lever or knob on the tub spout. Although gate diverters rarely need repair, the lever occasionally may break, come

loose, or refuse to stay in the UP position. To repair a gate diverter set in a tub spout, replace the entire spout (page 73).

Tub and shower faucets and diverter valves may be set inside wall cavities. Removing them may require a deep-set ratchet wrench (pages 71, 73).

If spray from the shower head is uneven, clean the spray holes. If the shower head does not stay in an upright position, remove the shower head and replace the O-ring (page 76).

To add a shower to an existing tub, install a flexible shower adapter. Several manufacturers make complete conversion kits that allow a shower to be installed in less than one hour.

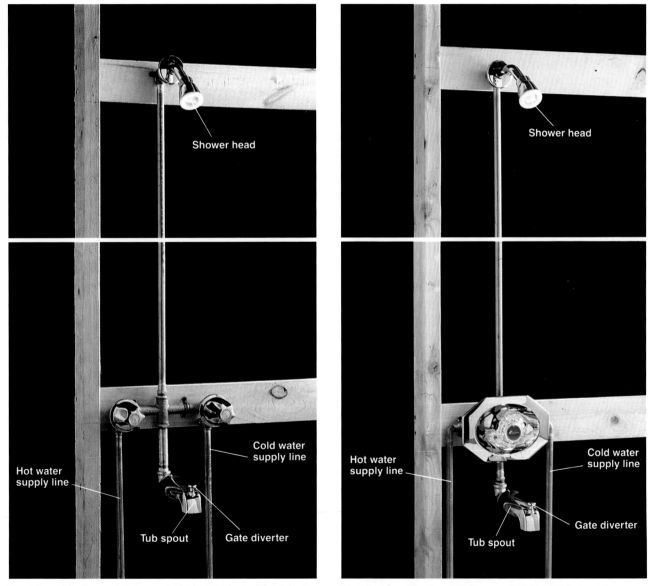

Two-handle faucet (pages 72 to 73) has valves that are either compression or cartridge design.

Single-handle faucet (pages 74 to 75) has valves that are cartridge, ball-type, or disc design.

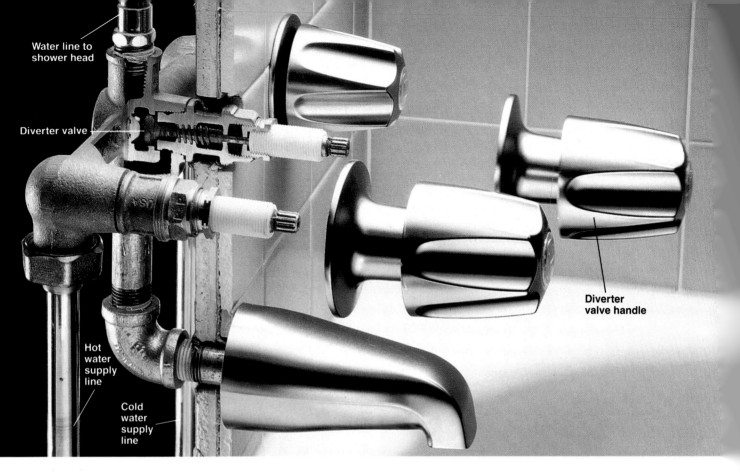

Water line to shower head

Diverter valve

Hot water supply line

Cold water supply line

Diverter valve handle

Fixing Three-handle Tub & Shower Faucets

A three-handle faucet type has handles to control hot and cold water, and a third handle that controls the diverter valve and directs water to either a tub spout or a shower head. The separate hot and cold handles indicate cartridge or compression faucet designs. To repair them, see pages 42 to 43 for cartridge, and 44 to 47 for compression.

If a diverter valve sticks, if water flow is weak, or if water runs out of the tub spout when the flow is directed to the shower head, the diverter needs to be repaired or replaced. Most diverter valves are similar to either compression or cartridge faucet valves. Compression type diverters can be repaired, but cartridge types should be replaced.

Remember to turn off the water (page 4) before beginning work.

Everything You Need:

Tools: screwdriver, adjustable wrench or channel-type pliers, deep-set ratchet wrench, small wire brush.

Materials: replacement diverter cartridge or universal washer kit, heatproof grease, vinegar.

How to Repair a Compression Diverter Valve

Escutcheon

Diverter valve handle

1 Remove the diverter valve handle with a screwdriver. Unscrew or pry off the escutcheon.

Bonnet nut

2 Remove bonnet nut with an adjustable wrench or channel-type pliers.

3 Unscrew the stem assembly, using a deep-set ratchet wrench. If necessary, chip away any mortar surrounding the bonnet nut (page 73, step 2).

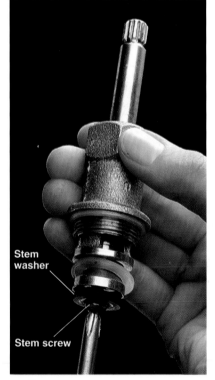

Stem washer

Stem screw

4 Remove brass stem screw. Replace stem washer with an exact duplicate. If stem screw is worn, replace it.

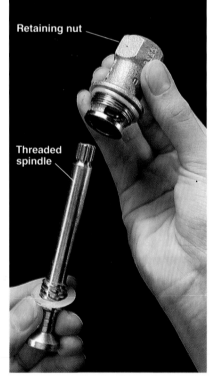

Retaining nut

Threaded spindle

5 Unscrew threaded spindle from retaining nut.

6 Clean sediment and lime build-up from nut, using a small wire brush dipped in vinegar. Coat all parts with heatproof grease and reassemble diverter valve.

Water line to
shower head

Bonnet
nut

Valve stem

Cold water
supply line

Hot water
supply line

Diverter lever

Gate diverter

Fixing Two-handle Tub & Shower Faucets

Two-handle tub and shower faucets are either cartridge or compression design. They may be repaired
following the directions on pages 42 to 43 for cartridge, or pages 44 to 47 for compression. Because
the valves of two-handle tub and shower faucets
may be set inside the wall cavity, a deep-set socket
wrench may be required to remove the valve stem.

Two-handle tub and shower designs have a gate
diverter. A gate diverter is a simple mechanism located in the tub spout. A gate diverter closes the
supply of water to the tub spout and redirects the
flow to the shower head. Gate diverters seldom need
repair. Occasionally, the lever may break, come
loose, or refuse to stay in the UP position.

If the diverter fails to work properly, replace the
tub spout. Tub spouts are inexpensive and easy
to replace.

**Remember to turn off the water (page 4) before
beginning work.**

Everything You Need:

Tools: screwdriver, allen wrench, pipe wrench,
channel-type pliers, small cold chisel, ball
peen hammer, deep-set ratchet wrench.

Materials: masking tape or cloth, pipe joint compound, replacement faucet parts as needed.

Tips on Replacing a Tub Spout

Spout nipple

Check underneath tub spout for a small access slot. The slot indicates the spout is held in place with an allen screw. Remove the screw, using an allen wrench. Spout will slide off.

Unscrew faucet spout. Use a pipe wrench, or insert a large screwdriver or hammer handle into the spout opening and turn spout counterclockwise.

Spread pipe joint compound on threads of spout nipple before replacing spout.

How to Remove a Deep-set Faucet Valve

Escutcheon

Masking tape

Stem nipple

Bonnet nut

1 Remove handle, and unscrew the escutcheon with channel-type pliers. Pad the jaws of the pliers with masking tape to prevent scratching the escutcheon.

2 Chip away any mortar surrounding the bonnet nut, using a ball peen hammer and a small cold chisel.

3 Unscrew the bonnet nut with a deep-set ratchet wrench. Remove the bonnet nut and stem from the faucet body.

Water supply line to shower head

Built-in shutoff valves

Control valve

Hot water supply line

Cold water supply line

Escutcheon

Gate diverter

Fixing Single-handle Tub & Shower Faucets

A single-handle tub and shower faucet has one valve that controls both water flow and temperature. Single-handle faucets may be ball-type, cartridge, or disc designs.

If a single-handle control valve leaks or does not function properly, disassemble the faucet, clean the valve, and replace any worn parts. Use the repair techniques described on pages 40 to 41 for ball-type, or pages 48 to 49 for ceramic disc. Repairing a single-handle cartridge faucet is shown on the opposite page.

Direction of the water flow to either the tub spout or the shower head is controlled by a gate diverter.

Gate diverters seldom need repair. Occasionally, the lever may break, come loose, or refuse to stay in the UP position. If the diverter fails to work properly, replace the tub spout (page 73).

Everything You Need:

Tools: screwdriver, adjustable wrench, channel-type pliers.

Materials: replacement parts as needed.

How to Repair a Single-handle Cartridge Tub & Shower Faucet

1 Use a screwdriver to remove the handle and escutcheon.

2 Turn off water supply at built-in shutoff valves or main shutoff valve (page 4).

3 Unscrew and remove retaining ring or bonnet nut, using an adjustable wrench.

4 Remove cartridge assembly by grasping end of valve with channel-type pliers and pulling gently.

5 Flush valve body with clean water to remove sediment. Replace any worn O-rings. Reinstall cartridge and test valve. If faucet fails to work properly, replace the cartridge.

- Shower arm
- Collar nut
- Swivel ball nut
- Spray adjustment cam lever
- Swivel ball
- O-ring
- Spray outlets

A typical shower head can be disassembled easily for cleaning and repair. Some shower heads include a spray adjustment cam lever that is used to change the force of the spray.

Fixing & Replacing Shower Heads

If spray from the shower head is uneven, clean the spray holes. The outlet or inlet holes of the shower head may get clogged with mineral deposits.

Shower heads pivot into different positions. If a shower head does not stay in position, or if it leaks, replace the O-ring that seals against the swivel ball.

A tub can be equipped with a shower by installing a flexible shower adapter kit. Complete kits are available at hardware stores and home centers.

Everything You Need:

Tools: adjustable wrench or channel-type pliers, pipe wrench, drill, glass & tile bit (if needed), mallet, screwdriver.

Materials: masking tape, thin wire (paper clip), heatproof grease, rag, replacement O-rings (if needed), masonry anchors, flexible shower adapter kit (optional).

How to Clean & Repair a Shower Head

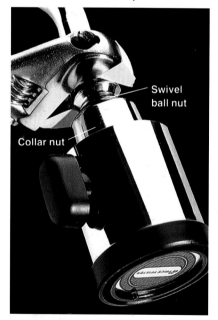

Swivel ball nut

Collar nut

1 Unscrew swivel ball nut, using an adjustable wrench or channel-type pliers. Wrap jaws of the tool with masking tape to prevent marring the finish. Unscrew collar nut from shower head.

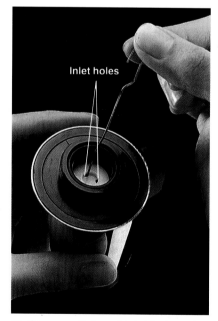

Inlet holes

2 Clean outlet and inlet holes of shower head with a thin wire. Flush the head with clean water.

O-ring

3 Replace the O-ring, if necessary. Lubricate the O-ring with heatproof grease before installing.

Fixing Burst or Frozen Pipes

When a pipe bursts, immediately turn off the water at the main shutoff valve. Make temporary repairs with a sleeve clamp repair kit (page opposite).

A burst pipe is usually caused by freezing water. Prevent freezes by insulating pipes that run in crawl spaces or other unheated areas.

Pipes that freeze, but do not burst, will block water flow to faucets or appliances. Frozen pipes are easily thawed, but determining the exact location of the blockage may be difficult. Leave blocked faucets or valves turned on. Trace supply pipes that lead to blocked faucet or valve, and look for places where the line runs close to exterior walls or unheated areas. Thaw pipes with a heat gun or hair dryer (below).

Old fittings or corroded pipe also may leak or rupture. Fix old pipes according to the guidelines described on pages 16 to 37.

Everything You Need:

Tools: heat gun or hair dryer, gloves, metal file, screwdriver.

Materials: pipe insulation, sleeve clamp repair kit.

Begin any emergency repair by turning off water supply at main shutoff valve. The main shutoff valve is usually located near water meter.

How to Repair Pipes Blocked with Ice

1 Thaw pipes with a heat gun or hair dryer. Use heat gun on low setting, and keep nozzle moving to prevent overheating pipes.

2 Let pipes cool, then insulate with sleeve-type foam insulation to prevent freezing. Use pipe insulation in crawl spaces or other unheated areas.

Alternate: Insulate pipes with fiberglass strip insulation and waterproof wrap. Wrap insulating strips loosely for best protection.

How to Temporarily Fix a Burst Pipe

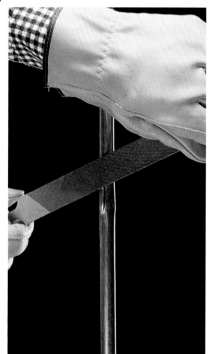

1 Turn off water at main shutoff valve. Heat pipe gently with heat gun or hair dryer. Keep nozzle moving. Once frozen area is thawed, allow pipe to drain.

2 Smooth rough edges of rupture with metal file.

3 Place rubber sleeve of repair clamp around rupture. Make sure seam of sleeve is on opposite side of pipe from rupture.

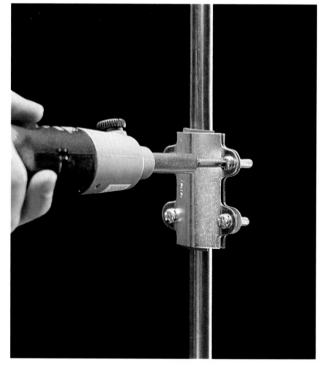

4 Place the two metal repair clamps around rubber sleeve.

5 Tighten screws with screwdriver. Open water supply and watch for leaks. If repair clamp leaks, retighten screws. **Caution: repairs made with a repair clamp kit are temporary.** Replace ruptured section of pipe as soon as possible.

Index

Cowles Creative Publishing, Inc.
offers a variety of how-to books.
For information write:
  Cowles Creative Publishing
  Subscriber Books
  5900 Green Oak Drive
  Minnetonka, MN 55343